New Techniques on Soft Soils

New Techniques on Soft Soils

Márcio Almeida

editor

Routledge
Taylor & Francis Group

LONDON AND NEW YORK

First published 2010 by Oficina de Textos

Published 2018 by Routledge
2 Park Square, Milton Park, Abingdon, Oxon OX14 4RN
52 Vanderbilt Avenue, New York, NY 10017

First issued in paperback 2018

Routledge is an imprint of the Taylor & Francis Group, an informa business

Lectures and keynote lectures from the Symposium on New Techniques for Design and Construction in Soft Clays, held on May 22 and 23, 2010, in Guarujá, Brazil.

Dados Internacionais de Catalogação na Publicação (CIP)
(Câmara Brasileira do Livro, SP, Brasil)

Symposium on New Techniques for Design and Construction in Soft Clays
(2010 : Guarujá, SP)

   New techniques on soft soils / Márcio Almeida, editor. – São Paulo : Oficina de Textos, 2010.

   ISBN 978-85-7975-002-1

   1. Argila 2. Aterros 3. Drenos verticais 4. Geologia de engenharia 5. Geotécnica 6. Mecânica do solo I. Almeida, Márcio. II. Título.

10-03775   CDD-624.15136

Índices para catálogo sistemático:

1. Simpósios : Anais : Engenharia geotécnica   624.15136

COVER AND GRAPHIC DESIGN Malu Vallim
LAYOUT Casa Editorial Maluhy & Co.
FIGURES PREPARATION Douglas da Rocha Yoshida

ISBN 13: 978-1-138-11200-1 (pbk)
ISBN 13: 978-85-7975-002-1 (hbk)

Preface

This volume is a compilation of the invited papers presented at the Symposium on New Techniques for Design and Construction in Soft Clays held in Guarujá/SP, Brazil, between 22 and 23 May 2010. This conference was organized by the Federal University of Rio de Janeiro, the Brazilian Society of Soil Mechanics (ABMS), and the Brazilian Chapter of the International Geosynthetics Society (IGS-Brazil), under the auspices of the International Society for Soil Mechanics and Geotechnical Engineering (ISSMGE) and the International Geosynthetics Society (IGS).

A number of techniques are presently available for site investigation, design and construction in soft clays. The range of the different techniques is quite wide and varies from country to country. These techniques are discussed in a number of papers presented in this book, with experiences and opinions of specialists from different countries.

The first section is about site investigation, vertical drains and surcharge, with a Keynote Lecture from Dr. Kerry Rowe concerning the interaction between drains and reinforcement and effect on performance of embankments on soft ground. Lectures on preloading design, vacuum consolidation, test embankment and *in situ* testing are also presented.

The second section deals with piled embankments, granular piles and deep mixing, with a Keynote Lecture from Dr. George Filz on deep mixing to improve the stability of embankments, levees, and floodwalls constructed on soft clay. Lectures also about numerical simulations, full-scale experiments, deep-vibro techniques and consolidation settlements are presented.

The third section is on the subject of monitoring and performance, with a Keynote Lecture from Dr. Márcio Almeida showing an overview of Brazilian practice of construction over soft soils and another from Dr. Buddhima Indraratna about soft soils improved by prefabricated vertical drains: performance and prediction. Lectures regarding the Brazilian experience in trial and pilot embankments, as well as in instrumented case histories are also presented.

In that way, the accumulated knowledge of professionals and researchers from 12 different countries are compiled on this volume providing an essential tool for designers and practitioners involved in soft soil construction.

MÁRCIO DE SOUZA S. ALMEIDA (Chairman)

Organized by

Alberto Luiz Coimbra Institute –
   Graduate School and Research in Engineering – COPPE-UFRJ
Brazilian Society for Soil Mechanics and Geotechnical Engineering - ABMS
Brazilian Association of Geosynthetics – IGS-Brazil

Under the Auspices of:

International Geosynthetics Society – IGS
International Society for Soil Mechanics and Geotechnical Engineering – ISSMGE

Scientific Committee:

José Renato M. S. Oliveira, IME, Brazil
Márcio de Souza S. Almeida, Chair, COPPE-UFRJ, Brazil
Marcos Barreto de Mendonça, Poli-UFRJ, Brazil
Marcos Massao Futai, EPUSP, Brazil
Maria Esther S. Marques, IME, Brazil
Maurício Ehrlich, COPPE-UFRJ, Brazil

Local Organizing Committee

André Pereira Lima, Geoinfra and UVA
Bruno Lima, COPPE-UFRJ and UFF
Celso Nogueira Corrêa, Zaclis Falconi
Diego de Freitas Fagundes, COPPE-UFRJ
José Renato Oliveira, IME
Khader Rammah, COPPE-UFRJ
Magnos Baroni, COPPE-UFRJ
Márcio de Souza S. Almeida, COPPE-UFRJ – Chair
Marcos Barreto de Mendonça, Poli-UFRJ
Marcos Massao Futai, EPUSP
Maria Esther S. Marques, IME
Mário Vicente Riccio Filho,COPPE-UFRJ

International Advisory Committee

Ennio M. Palmeira, University of Brasilia, Brazil
Jorge G, Zornberg, The University of Texas at Austin, USA
Osamu Kusakabe, Tokyo Institute of Technology, Japan
R. N. Taylor, City University, UK
Sarah M. Springman, ETH-Zurich, Switzerland
Serge Varaksin, France

Table of Contents

SECTION 3
Monitoring & Performance

KEYNOTE LECTURES

LECTURES

Section I | Site Investigation, Vertical Drains and Surcharge

Keynote lecture

The interaction between reinforcement and drains and their effect on the performance of embankments on soft ground

Rowe, R. K. and Taechakumthorn, C.
GeoEngineering Centre at Queen's-RMC
Department of Civil Engineering, Queen's University, Canada

KEYWORDS embankment, reinforcement, geosynthetics, PVDs, creep/relaxation, soft ground, finite element analysis, design methods

ABSTRACT This paper reviews the behaviour of reinforced embankments on both typical soft cohesive (rate-insensitive) soil and rate-sensitive soil. The interaction between geosynthetic reinforcement and prefabricated vertical drains (PVDs) is examined and it is demonstrated that the combination of PVDs and the tension mobilized in reinforcement can substantially increase the stability of embankments. The paper also provides a brief explanation of a recent design approach for embankments on soft rate-insensitive soil, considering the combined effect of reinforcement and PVDs. The effect of creep/relaxation of geosynthetics and rate sensitivity of the foundation soil on embankment performance is discussed and it is shown that they can have a significant effect on the failure height of reinforced embankments. The results suggest the need for considerable care when the foundation soil is rate-sensitive.

1 INTRODUCTION

Geosynthetics reinforcement and prefabricated vertical drains (PVDs) have revolutionized many aspects of the design and construction of embankments on soft ground and provide a cost effective alternative to traditional techniques. The behaviour of reinforced embankments on typical soft deposits is now well understood and many design procedures have been proposed. However, while these design methods may be conservative for conventional (rate-insensitive) soils, they may be quite unconservative for less conventional (rate-sensitive) soils (Rowe and Li, 2005; Li and Rowe, 2008 and Rowe and Taechakumthorn, 2008a).

The beneficial effects of PVDs for accelerating the gain in soil strength are well recognized (e.g. Taechakumthorn and Rowe, 2008; Sinha et al., 2009 and Saowapakpiboon et al., 2009). For example, when PVDs are used in conjunction with basal reinforcement, the presence of PVDs can substantially reduce the long-term creep deformation while allowing more rapid construction than could be safely considered without the use of PVDs (Li and Rowe, 2001 and Rowe and Taechakumthorn, 2008a).

Thus, the objective of this paper is to review research relating to the effects of the basal reinforcement and PVDs on the design and construction of embankments over soft ground. This paper also summarizes a design approach, which considers the interaction between reinforcements and PVDs, for embankments on typical soft clay deposits. A number of parametric studies are used to highlight some design considerations and potential problems that might be anticipated during the construction. This paper follows two previous keynote papers (Rowe and Li, 2005 and Rowe and Taechakumthorn, 2008b) and while it covers some of the same basic material as those papers, it also provides new insights based on the most recent research in the field up to November 2009.

2   REINFORCED EMBANKMENT ON SOFT GROUND

When embankments are constructed on soft foundations, the lateral earth pressure within the embankment fill imposes shear stresses on the foundation reducing its bearing capacity and hence embankment stability (Jewell, 1987). The role of the basal reinforcement is to absorb some, or all, of the earth pressure from the fill and to resist the lateral deformations of the foundation, thereby increasing embankment stability. Reinforced embankments are typically designed based on consideration of: (a) bearing capacity, (b) global stability, (c) pullout or anchorage, and (d) deformations (Leroueil and Rowe, 2001). The significant role that geosynthetic reinforcement can play in increasing embankment stability on traditional soft clay deposits was first clearly demonstrated from a theoretical perspective by the analysis of the Almere test embankments (Rowe and Soderman, 1984). They clearly demonstrated that the reinforcement does very little (and there are negligible strains) while the soil remains elastic. Strains begin to develop as the plastic zone grows and they develop rapidly when there is contiguous plastic failure in the soil and the reinforcement is all that is maintaining the embankments stability.

3   PARTIALLY DRAINED BEHAVIOUR OF REINFORCED EMBANKMENTS

The observed construction-induced excess pore water pressures from a large number of field cases suggest that partial consolidation of the foundation may occur during embankment construction at typical construction rates (Crooks et al., 1984; Leroueil and Rowe, 2001). This applies to natural soft cohesive deposits that are typically slightly overconsolidated. Also, it has been reported that there can be a significant strength gain due to partial consolidation during embankment construction (e.g. Chai et al., 2006 and Saowapakpiboon et al., 2009).

Although field cases suggest the importance of considering partial drainage, they do not allow a direct comparison of cases where it is and is not considered. Finite element analyses provide a powerful tool for comparing the behaviour of reinforced embankments constructed under both undrained and partially drained conditions (Rowe and Li, 2005).

When soft foundations do not initially have the strength to safely support a given embankment, stage construction may be employed to allow sufficient consolidation and strength gain to support the final embankment load. Li and Rowe (2001) showed that geosynthetic basal reinforcement may eliminate the need for stage construction or, in cases where staging was still needed, reduce the number of stages required. Their results also implied that there may be benefits arising from the combined use of reinforcement with methods of accelerating consolidation, such as PVDs, as discussed in the following section.

4   INTERACTION BETWEEN REINFORCEMENT AND PVDS

Since the first prototype of a prefabricated drain made of cardboard, PVDs have been widely used in embankment construction projects because of the benefits in terms of reduced construction costs and improved ease of construction (e.g. Holtz et al., 2001; Bergado et al., 2002; Chai et al., 2006; Sinha et al., 2009 and Saowapakpiboon et al., 2009). PVDs accelerate soil consolidation by shortening the drainage path and taking advantage of the naturally higher horizontal hydraulic conductivity of the foundation soil. This technique improves embankment stability by allowing strength gain in the foundation soil associated with the increase in effective stress due to consolidation.

The combined effects of reinforcement and PVDs have been investigated by Li and Rowe (2001) and Rowe and Taechakumthorn (2008a). It has been shown that at typical construction rates, the use of PVDs results in relatively rapid dissipation of excess pore pressures and this can be enhanced by the use of basal reinforcement. For example, Fig. 1 shows the variation of net embankment height (fill thickness minus maximum settlement) with embankment fill thickness from finite element simulations, where S is the spacing of PVDs in a square pattern. For this particular foundation soil (see insert in Fig. 1) and PVDs at a spacing of 2 m, the unreinforced embankment can be constructed to a height of 2.85 m. If reinforcement with tensile stiffness $J = 250$ kN/m is used, the failure height increases to 3.38 m. It is noted that, for these assumed soil properties and a construction rate of 2 m/month, failure of the embankment due to excessive settlement of the foundation soil will not occur if the reinforcement stiffness is greater than 500 kN/m.

When PVDs are used in conjunction with basal reinforcement, they can enhance the beneficial effect of the reinforcement in reducing the horizontal deformations of the foundation soil below the embankment as illustrated in Fig. 2. With the use of PVDs, less stiff reinforcement can be employed while

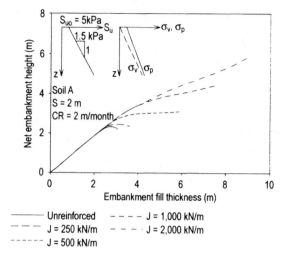

FIG. 1 *The combined effect of reinforcement and PVDs on the short-term stability of an embankment (modified from Rowe and Li, 2005)*

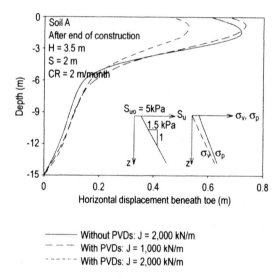

FIG. 2 *The combined effect of reinforcement and PVDs on lateral deformation beneath the toe of an embankment at the end of construction (modified from Rowe and Li, 2005)*

still providing about the same lateral deformation as when the stiffer reinforcement is used without PVDs.

5   DESIGN OF EMBANKMENTS ON SOFT GROUND CONSIDERING THE INTERACTION BETWEEN BASAL REINFORCEMENT AND PVDS

In design, the effects of reinforcement and PVDs are usually treated separately even if both are used. Li and Rowe (2001) proposed a design method for reinforced embankments that incorporates the effect of strength gain due to consolidation of the foundation soil. This design method was based on a limit state design philosophy using an undrained strength analysis suggested by Ladd (1991). The design procedure consists of four steps as described below.

First, the design criteria and representative soil parameters have to be selected including embankment geometry, soil profile (i.e. undrained shear strength, preconsolidation pressure, vertical effective stress, and coefficient of consolidation in both the normally consolidated and overconsolidated state, as well as vertical and horizontal hydraulic conductivity), longest vertical drainage path, embankment fill properties, and the anticipated average construction rate and hence the time available for consolidation. Second, one must select the configuration of the PVDs (pattern: triangular or square, spacing and length of PVDs). Then the method proposed by Li and Rowe (2001) can be utilized to calculate the average degree of consolidation at any specific time. Third, this information is used to calculate the increase in undrained shear strength along the potential failure surface using the SHANSEP method (Ladd, 1991) as described by Li and Rowe (2001). Finally, using this undrained shear strength, the limit equilibrium analysis can be employed to determine the reinforcement force required to give an adequate factor of safety and the required reinforcement stiffness is then calculated based on the magnitude of the required force and the allowable reinforcement strain. This then allows the selection of the required geosynthetic reinforcement. If reinforcement having the required stiffness and strength is not readily available, the process can be repeated deducing the spacing between the PVDs until a suitable combination of PVDs and reinforcement is identified. Full details of this approach and an example associated with the design approach summarized are provided by Li and Rowe (2001).

This approach can be easily applied for a stage construction sequence by adding the consolidation during the stoppage between stages when calculating the average degree of consolidation, while keep the other steps the same. To ensure embankment stability during construction, it is important to monitor the development of reinforcement strains, excess pore pressure, settlement, and horizontal deformation to confirm that the observed behaviour is consistence with the design assumptions (Rowe and Li, 2005).

6   REINFORCED EMBANKMENT ON RATE-SENSITIVE SOIL

It has been recognized by many researchers (Lo and Morin, 1972; Vaid and Campanella, 1977; Vaid et al., 1979; Graham, 1983; and Leroueil, 1988) that natural soft deposits exhibit significant time-dependent behaviour and the undrained shear strength of the natural soft clay is strain rate dependent (rate-sensitive). The performance of reinforced embankment constructed on the rate-sensitive soil also has been investigated by both field studies and numerical analysis (Rowe et al., 1996; Hinchberger and Rowe, 1998; Rowe and Hinchberger, 1998; Rowe and Li, 2002; and Rowe and Taechakumthorn, 2008a,b). Rowe et al. (1996) showed that in order to accurately predict the responses of the Sackville embankment on a rate-sensitive soil, it was essential to consider the effect of soil viscosity. Rowe and Hinchberger (1998) proposed an elasto-viscoplastic constitutive model based on the concept of overstress viscoplasticity using fluidity parameters to model the effect of soil viscosity and demonstrated that the model could adequately describe the behaviour of the Sackville test embankment.

For rate-sensitive soils, the undrained shear strength is highly dependant on the rate of loading (i.e. rate of embankment construction); the faster the loading rate, the stronger the soil appears. For that reason, the loading rate is an important factor when conducting an analysis of embankment performance on a rate-sensitive soil. The effect of construction rate and geosynthetic reinforcement on the short-term stability of reinforced embankments is illustrated in Fig. 3. Series of reinforced embankments with axial stiffness of 0 (unreinforced), 500 and 750 kN/m were numerically constructed at different construction rates until failure. The effect of construction rate is evident with the faster construction rate resulting in a higher short-term embankment failure height for all cases. Also, the use of basal reinforcement has been shown to improve embankment stability. Until a limiting stiffness is reached, the stiffer the reinforcement the higher confining force to the system and hence the higher the short-term failure height (Fig. 3). However, this short-term benefit hides a long-term problem as will be discussed later.

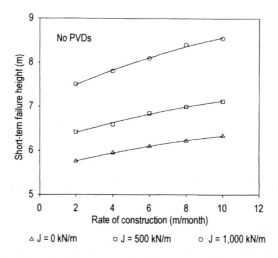

FIG. 3 *The effect of construction rate and reinforcement stiffness on short-term stability of an embankment (modified from Rowe and Taechakumthorn, 2008a)*

To investigate the effect of the various parameters such as reinforcement stiffness, construction rate and the effect of PVDs on the long-term behaviour of reinforced embankments on rate-sensitive soil, a series of 5 m high reinforced embankments were numerically constructed on rate-sensitive foundation soil A. The results from Case I and Case II (Fig. 4) show the effect of construction rate. The reinforcement strains at the end of the construction were 1.6% and 2.6% for Cases I and II, respectively. The reinforcement strain for the slower construction rate (Case II) was higher because the soil exhibited lower short-term strength and transferred more load to the geosynthetic reinforcement. However, this slower construction rate allowed a higher degree of partial consolidation and reduced the amount of overstress in the soil. Consequently, there was less creep and stress relaxation in the soil following construction. This resulted in smaller long-term reinforcement strains. The results from Case I and III (Fig. 4) show the effect of reinforcement stiffness and as expected the stiffer reinforcement (Case III) gave smaller strains both at the end of construction and long-term.

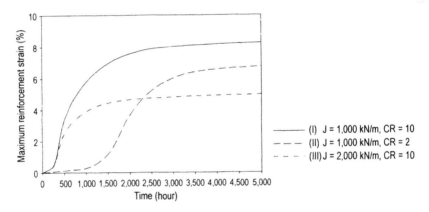

FIG. 4 *The effect of construction rate and reinforcement stiffness on mobilized reinforcement strains (modified from Rowe and Taechakumthorn, 2008a)*

The rate of excess pore water dissipation and the consequent rate of shear strength gain in the soil can be increased using PVDs. Results in Fig. 5 show that with the use of PVDs, the long-term mobilized reinforcement strain can be significantly reduced. For the 5 m high reinforced embankment with the reinforcement stiffness J = 1,000 kN/m, even a construction rate as low as 2 m/month gave rise to a long-term reinforcement strain of 6.9% which exceeds the typical allowable limit of about 5% (Fig. 4). However, with PVDs at 3 m spacing, numerical construction of the embankment to the same h = 5 m height at 10 m/month gave a maximum long-term reinforcement strain of 4.6% (Case I, Fig. 5). With stiffer (J = 2,000 kN/m) reinforcement, PVDs reduced the long-term reinforcement strain from 4.9% to 3.3% (Case III in Fig. 4 and Case

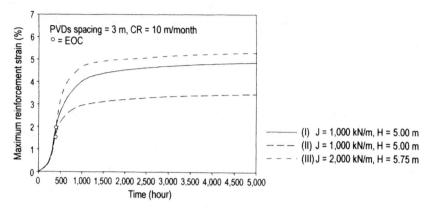

FIG. 5 *The effect of PVDs and reinforcement stiffness on mobilized reinforcement strains (modified from Taechakumthorn and Rowe, 2008)*

II, Fig. 5). With a reinforcement stiffness of 2,000 kN/m, a reinforced embankment could be constructed up to 5.75 m without the long-term reinforcement strain exceeding about 5% (Case III, Fig. 5).

7   EFFECTS OF CREEP/RELAXATION OF REINFORCEMENT STAINS

Experimental studies have shown that geosynthetics typically made of polyester (PET), polypropylene (PP) and polyethylene (PE) are susceptible to creep/relaxation (Jewell and Greenwood, 1988; Greenwood, 1990; Bathurst and Cai, 1994; Leshchinsky et al., 1997; Shinoda and Bathurst, 2004; Jones and Clarke, 2007; and Kongkitkul and Tatsuoka, 2007). The importance of considering creep/relaxation of geosynthetics reinforcement, to understand the time-dependent behaviour of the reinforced embankments on soft ground has been highlighted in the literature (Rowe and Li, 2005; Li and Rowe, 2008; and Rowe and Taechakumthorn, 2008b).

For creep-sensitive reinforcement, in some cases the reinforcement strain may significantly increase with time after embankment construction (Rowe and Li, 2005). Fig. 6 shows (solid lines) the development of reinforcement strain with time up to 98% consolidation for embankments reinforced with high density polyethylene, HDPE, (upper figure) and PET (lower figure) geosynthetics. Also shown (dashed lines) are the strains that would be developed if the reinforcement was assumed elastic with stiffness selected such that, at the end of construction, the reinforcement strain is the same as that developed in the viscous reinforcement. Thus, the difference between the solid and dashed lines represents the creep strain due to the viscous nature of the reinforcement.

Rowe and Li (2005) demonstrated that the isochronous stiffness deduced from standard creep tests can reasonably represent the stiffness of

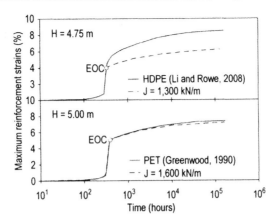

FIG. 6 *Variation of reinforcement strain with time during and following the embankment construction over a rate-insensitive soil (modified from Rowe and Li, 2005)*

geosynthetics reinforcement at the critical stage, for rate-insensitive foundation soils. The study also recommended that the isochronous stiffness should be used in design to estimate the mobilized reinforcing force at the end of embankment construction. Fig. 7 compares the mobilized reinforcement stiffness with isochronous stiffness deduced from in-isolation creep test data during and after the construction of the HDPE geogrid and PET geosynthetic-reinforced embankments. The mobilized stiffness decreases with time and very closely approaches the isochronous stiffness in the long term. This also agrees with the finding of Li and Rowe (2008) and Rowe and Taechakumthorn (2008b) for the case of rate-sensitive foundation.

The time-dependence of the mobilized reinforcement stiffness shown in Fig. 7 also implies that the force in the reinforcement following the end

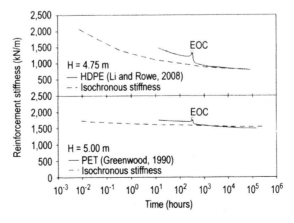

FIG. 7 *Variation of reinforcement tensile stiffness with time during and following the embankment construction (modified from Rowe and Li, 2005)*

of embankment construction may be significantly lower than expected in design owing to the viscous behaviour of geosynthetic reinforcement during embankment construction. This highlights the need for care when applying tensile stiffness from standard load–strain tests to deduce the design tensile force. In addition to creep effects, consideration should be given to potential effects of construction damage to the reinforcement.

8   Conclusions

The behaviour of reinforced embankments and the current design approaches have been examined for a number of different situations. The results show that the use of geosynthetic reinforcement can substantially increase the failure height of embankments over soft ground. The finite element method has proven to be an effective tool to analyze the behaviour of reinforced embankments. The results showed that the performance of the reinforced embankment can change significantly, depending on the type of geosynthetic used and/or the nature of the foundation soil. Therefore, careful consideration must be given to the selection of the constitutive relationships to model each component of a reinforced embankment. Basal reinforcement can improve the stability of the embankment on both conventional (rate-insensitive) as well as rate-sensitive soil. Furthermore, the effect of partial consolidation during embankment construction can enhance the effect of reinforcement, which encourages the combined use of reinforcement, with methods of accelerating consolidation, such as PVDs. When stage construction is required, the use of reinforcement can reduce the number of stages needed. With the presence of PVDs, the design method proposed by Li and Rowe (2001) can be employed to address the interaction between the effect of strength gain, associated with the partial consolidation, and the reinforcement.

For reinforced embankments constructed over rate-sensitive soil, although the viscoplastic nature of the foundation can increase the short-term stability of the embankment, it significantly degrades the long-term embankment stability following the end of construction. The use of reinforcement provides a confining stress to the system and limits creep in the foundation. PVDs can provide a significant enhancement to the performance of reinforced embankments on these soils. For example, because PVDs allow a higher degree of consolidation during and following the construction, the overstress and consequent creep in the soil is reduced, resulting in less differential settlement and lateral movement as well as smaller long-term reinforcement strains.

As a result of the time-dependent nature of the geosynthetic reinforcement, reinforcement stiffness at the end of construction is less than that provided by the standard tensile test. This implies that the reinforcement force used in the design may not represent what has been mobilized in the field. The isochronous stiffness measured from a standard creep tests appears to reasonably, and conservatively, represent the reinforcement stiffness in the field at the end of construction. The results also suggest that reinforcement creep and stress-relaxation allow an increase in the shear deformations of the foundation soil which will degrade the long-term performance of the reinforced embankment and may even lead to long-term failure, if the foundation soil is rate sensitive. Considerable care must be taken in the design of embankments when dealing with both creep-susceptible reinforcement and a rate-sensitive foundation soil.

ACKNOWLEDGEMENTS

All the work reported in this paper was funded by the National Sciences and Engineering Research Council of Canada.

REFERENCES

Bathurst, R. J. and Cai, Z. (1994). In-isolation cyclic load-extension behavior of two geogrids. *Geosynthetics International*, Vol. 1(1), pp. 3–17.

Bergado D. T., Balasubramaniam, A. S., Fannin, R. J., Anderson, L. R. and Holtz, R. D. (2002). Prefabricated vertical drains (PVDs) in soft Bangkok clay: a case study of the new Bangkok International Airport project. *Canadian Geotechnical Journal*. Vol. 39, pp. 304-315.

Chai, J. C., Carter, J. P. and Hayashi, S. (2006). Vacuum consolidation and its combination with embankment loading. *Canadian Geotechnical Journal*. Vol. 43, pp. 985-996.

Crooks, J. H., Becker, D. E., Jeffries, M. G. and McKenzis, K. (1984). Yield behaviour and consolidation I: Pore pressure response. *Proc. ASCE Symposium on Sedimentation Consolidation Models*, Prediction and Validation, San Francisco, pp. 356-381.

Graham, J., Crooks, H. A. and Bell, A. L. (1983). Time effects on stress-strain behaviour of natural soft clays. *Geotechnique*, Vol. 33, pp. 327-340.

Greenwood, J. H. (1990). The creep of geotextiles. *Proceedings 4th International Conference on Geotextiles and Geomembranes and Related Products*, The Hague, Vol. 2, pp. 645-650.

Hinchberger, S. D. and Rowe, R. K. (1998). Modelling the rate-sensitive characteristics of the Gloucester foundation soil. *Canadian Geotechnical Journal*, Vol. 35, pp. 769–789.

Holtz, R. D., Shang, J. Q. and Bergado, D. T. (2001). Soil improvement. *In Geotechnical and Geoenvironmental Engineering Handbook* (ed. R. K. Rowe). Norwell, MA: Kluwer Academic, pp. 429-462.

Jewell, R. A. and Greenwood, J. H. (1988). Long-term strength and safety in steep soil slopes reinforced by polymer materials. *Geotextiles and Geomembranes*, Vol., 7, pp. 81-118.

Jones, C. J. F. P. and Clarke, D. (2007). The residual strength of geosynthetic reinforcement subjected to accelerated creep testing and simulated seismic events, *Geotextiles and Geomembranes*, Vol. 25, pp. 155-169.

Kongkitkul, W. and Tatsuoka, F. (2007). A theoretical framework to analyse the behaviour of polymer geosynthetic reinforcement in temperature-accelerated creep test. *Geosynthetics International*, Vol. 14(1), pp. 23-38.

Ladd, C. C. (1991). Stability evaluation during staged construction. *Journal of Geotechnical Engineering*, Vol. 117(4), pp. 540-615.

Leroueil, S. (1988). Tenth Canadian Geotechnical Colloquium: Recent developments in consolidation of natural clays. *Canadian Geotechnical Journal*, Vol., 25(1), pp. 85-107.

Leroueil, S. and Rowe, R. K. (2001). Embankments over soft soil and peat. *In Geotechnical and Geoenvironmental Engineering Handbook* (ed. R. K. Rowe). Norwell, MA: Kluwer Academic, pp. 463-500.

Leshchinsky, D., Dechasakulsom, M., Kaliakin, V. N. and Ling, H. I. (1997). Creep and stress relaxation of geogrids. *Geosynthetics International*, Vol. 4(5), pp. 463-479.

Li, A. L. and Rowe, R. K. (2001). Combined effects of reinforcement and pre-fabricated vertical drains on embankment performance. *Canadian Geotechnical Journal*, Vol. 38, pp.1266-1282.

Li, A. L. and Rowe, R. K. (2008). Effects of viscous behaviour of geosynthetic reinforcement and foundation soils on the performance of reinforced embankments. *Geotextiles and Geomembranes*, Vol. 26, pp. 317-334.

Lo, K. Y. and Morin, J. P. (1972). Strength anisotropy and time effects of two sensitive clays. *Canadian Geotechnical Journal*, Vol. 9, pp.261-277.

Rowe, R. K. and Taechakumthorn, C. (2008a). Combined effect of PVDs and reinforcement on embankments over rate-sensitive soils. *Geotextiles and Geomembranes*, Vol. 26(3), pp. 239-249.

Rowe, R. K. and Taechakumthorn, C. (2008b). The effect of soil viscosity on the behaviour of reinforced embankment, Keynote lecture, *1st International Conference on Transportation Geotechnics*. Nottingham, UK, pp. 47-56.

Rowe, R. K and Hinchberger, S. D. (1998). The significance of rate effects in modelling the Sackville test embankment. *Canadian Geotechnical Journal*, Vol. 33, pp. 500–516.

Rowe, R. K. and Li, A. L. (2002). Behaviour of reinforced embankments on soft rate sensitive soils. *Geotechnique*, Vol. 52(1), pp. 29-40.

Rowe, R. K. and Li, A. L. (2005). Geosynthetic-reinforced embankments over soft foundations. *Geosynthetics International*, Vol. 12(1), pp. 50-85.

Rowe, R. K. and Soderman, K. L. (1984). Comparison of predicted and observed behaviour of two test embankments. *Geotextiles and Geomembranes*, Vol. 1(2), pp. 143-160.

Rowe, R. K., Gnanendran, C. T., Landva, A. O. and Valsangkar, A. J. (1996). Calculated and observed behaviour of a reinforced embankment over soft compressible soil. *Canadian Geotechnical Journal*, Vol. 33, pp. 324–338.

Saowapakpiboon, J., Bergado, D. T., Youwai, S., Chai, J. C., Wanthong, P. and Voottipruex, P. (2009). Measured and predicted performance of prefabricated vertical drains (PVDs) with and without vacuum preloading, *Geotextiles and Geomembranes*, In Press.

Shinoda, M. and Bathurst, R. J. (2004). Lateral and axial deformation of PP, HDPE and PET geogrids under tensile load. *Geotextiles and Geomembranes*, Vol. 22, pp. 205–222.

Sinha, A. K., Havanagi, V. G. and Mathur, S. (2009). An approach to shorten the construction period of high embankment on soft soil improved with PVD, *Geotextiles and Geomembranes*, Vol. 27(6), pp. 488-492.

Taechakumthorn, C. and Rowe, R. K. (2008). Behaviour of reinforced embankments on rate-sensitive foundation installed with prefabrication vertical drains, *12th International Conference of International Association for Computer Methods and Advances in Geomechanics (IACMAG)*, Goa, India, 3559-3566.

Vaid, Y. P. and Campanella, G. (1977). Time-dependent behaviour of undisturbed clay. *Journal of the Geotechnical Engineering*, Vol. 103(GT7), pp. 693-709.

Vaid, Y. P., Robertson, P. K. and Campanella, R. G. (1979). Strain rate behaviour of Saint-Jean-Vianney clay. *Canadian Geotechnical Journal*, Vol. 16, pp. 34–42.

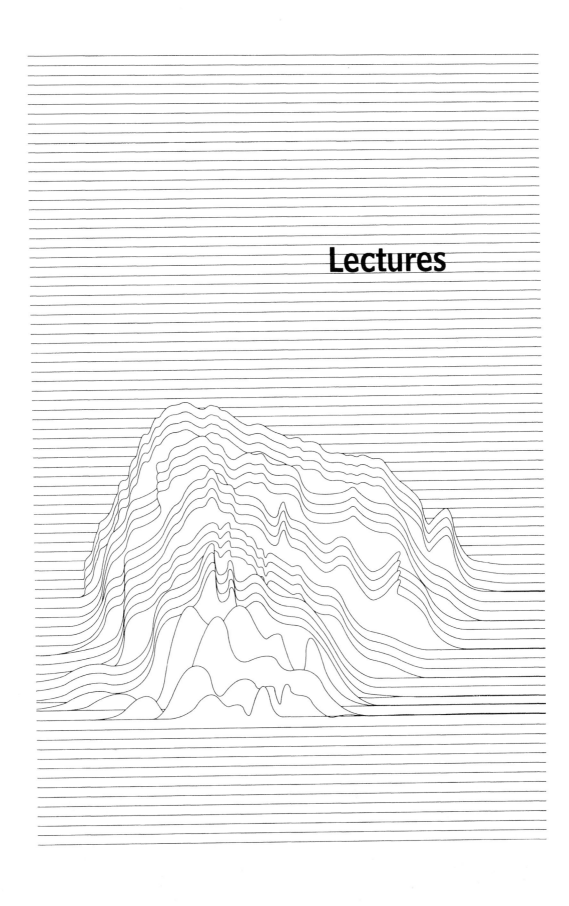

Lectures

Preloading Design Based on Long Term Extensometer Readings. A Comparison of Alternative Models

Alonso, E. E., Gens, A., Madrid, R. and Tarragó, D.
Department of Geotechnical Engineering and Geosciences. UPC, Barcelona

KEYWORDS Soft soil, settlement, creep, preloading, *in situ* test, models, consolidation

ABSTRACT The paper describes the analysis of a large scale preloading test aimed at determining precisely soft soil properties in a deltaic area. Deep continuous extensometers provided the key information to validate the settlement models. Two alternatives: an elastoplastic 2D plane strain FE model and one dimensional consolidation model were used to reproduce the long term deformation records, including an unloading stage. The two models were then used to predict the behaviour of the soil under the project service loading.

1 THE PROBLEM

The construction of new container terminals at the Barcelona Harbour requires the improvement of deep soft deltaic soils in order to limit future settlements under the design loads.

Preloading was selected as a convenient soil improvement procedure.

Preloading design is, at first sight, a relatively simple task from a conceptual point of view. Two main decisions are typically required: deciding the intensity of preloading and its duration.

Difficulties arise in practice because of the limitations to determine precisely the geometry of the consolidating foundation soils. The nature of internal drainage, the *in situ* primary and secondary deformability and the development in time of settlements after preloading are difficult to establish on the basis of conventional soil investigation procedures. As a further example of difficulties encountered, the secondary compression rate is controlled by the overconsolidation ratio (OCR), a variable changing in space and time, which depends on a number of aspects: the initial OCR profile, the actual stress distribution and the preloading times.

Alonso et al. (2000) described the capabilities of a well instrumented preloading test to investigate the layering of deep deltaic deposits and to determinate the variation of secondary settlement rates with the OCR. The work reported here elaborates further some of the previous findings, presents some improvements of the one dimensional consolidation model developed previously and specially compares the performance of the 1D model and a more involved visco-elasto-plastic finite element analysis. The new experience shows the definite advantage of a well instrumented preloading test over conventional soil investigation techniques.

The preloading test occupied a 160 m × 80 m rectangular area (Fig. 1). Sliding Micrometers provided an accurate record of vertical deformations at 1 m intervals. Piezometers are of the vibrating wire type and offered time records of pore pressures at some depths during the loading, consolidation and unloading stages. Both types of instruments could be directly compared with model predictions.

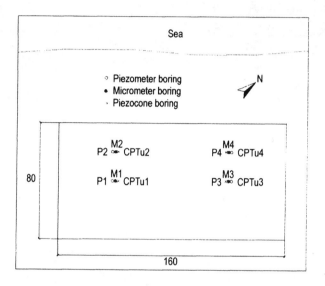

FIG. 1 *Preloaded area (180 m × 60 m). Location of CPTU tests and borings for continuous extensometers and piezometers*

Micrometers were also fundamental to identify precisely the detailed soil layering in terms of its stiffness. Therefore, they provide additional information to the data derived from Cone Penetration Tests with pore pressure measurements, CPTU's.

The soil profile is, in general terms, described by a sequence of fine sands, silts, and clays as shown in Fig. 2a. A granular substratum provides a stiff lower boundary. CPTU tests provided a detailed picture of layering (Fig. 2b), which is only approximately correlated with the visual description of the stratigraphic sequence.

Clayey levels were classified as CL (low plasticity clay) or ML (low plasticity silt), following the Unified Soil Classification System. Liquid limits and plasticity indices remain in the range 34-36% and 13-17% respectively. However, a significant proportion of the profile is classified as silts, sandy silts and sands of no plasticity.

The void ratio measured in recovered samples is quite variable. Fine clayey and silt materials reach values of e = 0.85−1.1, but there are also denser sandy levels (e = 0.5 − 0.7). Measured confined virgin compression coeffi-

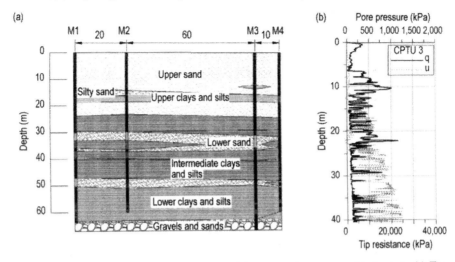

FIG. 2 *a) Statigraphic soil profile derived from samples recovered in borings. b) Tip resistance and pore pressure response of CPTU 3*

cients (oedometer tests) for the softer layers lie in the range $C_c = 0.1 - 0.2$. In sandy layers this coefficient may reduce to $C_c = 0.05$. Measured coefficients of consolidation span a wide range (8.10^{-2} cm^2/s for sandy soils to 5.10^{-4} cm^2/s for the more plastic ones). Measured secondary compression rates were uniform for all the samples tested (14 tests): $C_\alpha = 1.7$ to 5.8×10^{-3}.

The summarized data were difficult to integrate into an accurate deformation model of the foundation soils. The instrumented preloading test offered better information.

Table 1 provides a summary of the geotechnical properties of the foundation soils.

TABLE 1 GEOTECHNICAL PROPERTIES OF FOUNDATION SOILS

Layers	UCS	LL (%)	PI (%)	e	C_c	C_s	c_v (cm^2/s)	C_a
Clayey soils	CL – ML	34 – 36	13 – 17	0.85 – 1.1	0.1 – 0.2	0.005 – 0.021	5.10^{-4}	$1.88 \cdot 10^{-3}$ – $5.1 \cdot 10^{-3}$
Sandy soils	SP – SM	—	—	0.5 – 0.7	0.05	0.004 – 0.008	8.10^{-2}	$1.7 \cdot 10^{-3}$ – $3.7 \cdot 10^{-3}$

2   INSTRUMENTED PRELOADING

The test embankment reached a maximum elevation of 10.5 m over mean sea level. A reference time t = 0 was established as the origin of instrument readings (28.10.05). Most of the embankment loading was applied in the interval t = 60 to 170 days. Then the embankment remained at full height until it was unloaded to h = 6.50 m in the period t = 370 – 390 days and a further lowering to h = 2.50 m from t = 460 to t = 480 days. Fig. 3 shows the sequence of loading, consolidation and unloading.

One example of micrometer readings during the loading period is given in Fig. 4. The plot shows the sequence of highly deformable layers in the first 30-35 m. Maximum measured deformations reached 45 mm/m (4.5%). The plot shows also a progressive stiffening of the soil profile below a depth of 35 m. This is attributed to the overconsolidation induced by past lowering of the piezometric head in the lower pervious gravels.

The integration of micrometer deformations from an assumed fixed point at a depth of 60 m provides a settlement record during the loading period. The four installed micrometers led to similar settlements – time plots, as shown in Fig. 5.

Maximum settlements reached 0.65 m. Independent topographic surveys provided similar results, which is an indication of the stiff character of foundations soils below the depth of 65 m.

Unloading the embankment in two steps resulted in a rebound measured also by the micrometers (Fig. 6).

Micrometers delivered also time records of relative displacements for 'virtual' layers 1 m thick extending from the soil surface to the lower gravels. They were most useful to create and validate soil deformation models as illustrated below.

Piezometers reacted also to loading-unloading in the manner shown in Fig. 7 for the three piezometers installed in borehole P3. The plot shows the excess pore pressure over the hydrostatic value. Note the strong pressure deficit recorded at depth at the position of the gravel layer. Also, some of the temporary increase in pore pressures, at constant embankment preloading, is attributed to other earth loading operations taking place in the vicinity of the preloading test.

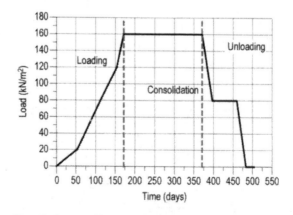

FIG. 3 *Stages of loading, consolidation and unloading*

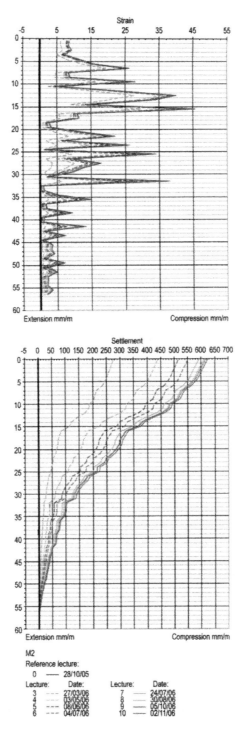

FIG. 4 *Time evolution of ground incremental and accumulated deformations in Micrometer M2 in the period 26/10/05-02/11/06 (370 days)*

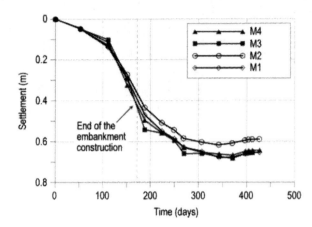

FIG. 5 *Time evolution of settlements during loading and consolidation based on strain measurements on micrometers M1, M2, M3 and M4 (natural time scale)*

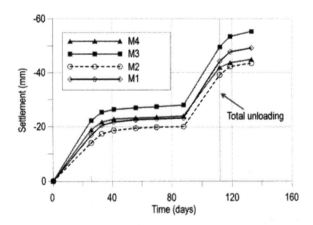

FIG. 6 *Time evolution of ground heave due to embankment unloading (based on strain measurements on micrometers M1, M2, M· and M4 (natural scale)*

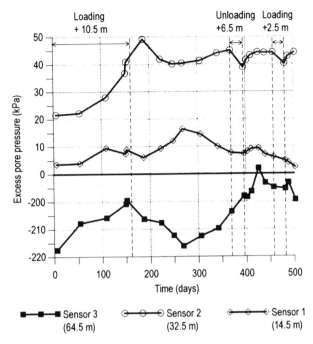

FIG. 7 *Time readings in sensors 1, 2 and 3 in piezometer borehole P3*

3 LAYERING GEOMETRY AND SOIL PARAMETERS

3.1 Two-dimensional plane strain elastoplastic modeling

The embankment material was represented by a conventional Coulomb material. A "soft soil creep" model, as defined in Plaxis, was selected for the natural soil layers. Soil deformation has two components: elastic (following modified Cam clay) and viscoplastic. The viscoplastic component requires the definition of a yield locus, which is made of a "cap" and a Mohr-Coulomb surface. The viscoplastic component is expressed in terms of a flow parameter directly related to the conventional secondary compression coefficient C_α. Details of this model may be found in the scientific manuals of Plaxis program, which is a general purpose finite element code for geotechnical analysis (Plaxis, 2010). Table 2 provides the list of model parameters.

A symmetric geometry with respect to the embankment major axis was defined. Only half of the full section was discretized as shown in Fig. 8. The soil layering was quite precise and followed the information provided by the four sliding micrometers. Layer thickness varied between 5 m and a minimum value of 1 m.

The process of parameter identification was a trial-and-error procedure. Starting at a given distribution of parameters, guided by the available soil description, the CPTU and oedometer tests and the micrometer data, the strains

TABLE 2 PARAMETERS OF THE MODEL "SOFT SOIL CREEP"

Symbol	Description	Units
γ_{dry}	Dry specific weight	kN/m^3
γ_{wet}	Saturated specific weight	kN/m^3
k_x	Horizontal permeability (principal component of tensor)	m/day
k_y	Vertical permeability (principal component of tensor)	m/day
$\lambda*$	Modified index of compressibility	—
$\kappa*$	Modified swelling index	—
$\mu*$	Modified flow index	—
c	Cohesion	kN/m^2
φ	Internal friction angle	°
ψ	Dilatancy angle	°
ν_{ur}	Poisson coefficient for elastic loading/unloading	—
M	Slope of critical state line (defines the "cap")	—
K_0^{NC}	At rest earth pressure coefficient (normally consolidated)	—

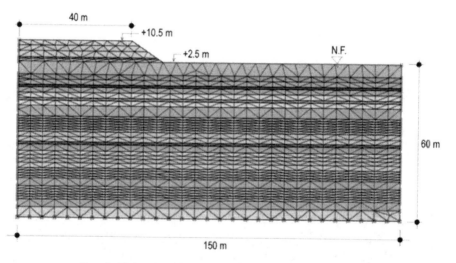

FIG. 8 *Finite element mesh for 2D plane strain analysis*

measured at regular intervals within the soil were progressively matched. Table 3 shows the parameters assigned to the successive clayey and sandy layers of the finite element model.

An important 'boundary' condition was given by the surface settlement records. Fig. 9 provides the matching achieved between incremental and accumulated displacements and the model calculations for micrometer M-2 for one particular time (t = 224 days).

The analysis followed also the actual estimated history of water level changes in the foundation soils. It is known that past intense pumping of the lower gravel aquifers induced a reduction of water heads in the order of 18 m. The hydrostatic water pressure distribution was accordingly modified

TABLE 3 SOFT SOIL-CREEP MODEL PARAMETERS USED CONSIDERED IN PLAXIS ANALYSIS

Material	γ_{dry} (kN/m³)	γ_{wet} (kN/m³)	k_x (cm/s)	k_y (cm/s)	λ^*	κ^*	μ^*	c' (kN/m²)	φ' (°)	ψ (°)
clay 1	19.0	19.0	$1.5.10^{-5}$	$7.5.10^{-6}$	0.050	$5.0.10^{-3}$	$6.0.10^{-4}$	1	27	0
clay 2	19.0	19.0	$1.5.10^{-5}$	$7.5.10^{-6}$	0.080	$4.0.10^{-3}$	$1.5.10^{-3}$	1	27	0
clay 3	19.0	19.0	$9.3.10^{-8}$	$2.5.10^{-6}$	0.058	$4.0.10^{-3}$	$1.2.10^{-3}$	1	27	0
clay 4	19.0	19.0	$1.6.10^{-7}$	$2.5.10^{-6}$	0.042	$2.5.10^{-3}$	$5.0.10^{-4}$	1	27	0
clay 5	19.0	19.0	$5.0.10^{-7}$	$1.0.10^{-5}$	0.063	$4.0.10^{-3}$	$1.0.10^{-3}$	1	27	0
clay 6	19.0	19.0	$1.6.10^{-7}$	$5.0.10^{-6}$	0.095	$6.0.10^{-3}$	$3.0.10^{-3}$	1	27	0
clay 7	19.0	19.0	$9.3.10^{-8}$	$2.5.10^{-6}$	0.046	$5.0.10^{-3}$	$1.0.10^{-3}$	1	27	0
clay 8	19.0	19.0	$6.9.10^{-8}$	$2.5.10^{-7}$	0.016	$2.0.10^{-3}$	$6.0.10^{-4}$	1	27	0
clay 9	19.0	19.0	$6.9.10^{-8}$	$1.0.10^{-6}$	0.043	$2.5.10^{-3}$	$8.0.10^{-4}$	1	27	0
clay 10	19.0	19.0	$6.9.10^{-8}$	$1.0.10^{-6}$	0.039	$2.0.10^{-3}$	$8.0.10^{-4}$	1	27	0
clay 11	19.0	19.0	$6.9.10^{-8}$	$1.0.10^{-6}$	0.048	$5.0.10^{-3}$	$8.0.10^{-4}$	1	27	0
clay 12	19.0	19.0	$6.9.10^{-8}$	$5.0.10^{-7}$	0.032	$1.5.10^{-3}$	$7.0.10^{-4}$	1	27	0
sand 1	20.0	20.0	$1.2.10^{-4}$	$1.2.10^{-4}$	0.007	$2.0.10^{-3}$	$2.0.10^{-4}$	1	32	0
sand 2	20.0	20.0	$1.2.10^{-4}$	$5.8.10^{-5}$	0.006	$1.1.10^{-3}$	$1.8.10^{-4}$	1	32	0
sand 2B	20.0	20.0	$1.2.10^{-4}$	$2.3.10^{-5}$	0.012	$1.5.10^{-3}$	$4.0.10^{-4}$	1	32	0
sand 2C	20.0	20.0	$1.2.10^{-4}$	$8.1.10^{-6}$	0.015	$1.5.10^{-3}$	$4.0.10^{-4}$	1	32	0
sand 3	20.0	20.0	$1.0.10^{-6}$	$5.0.10^{-7}$	0.025	$4.0.10^{-3}$	$9.0.10^{-4}$	1	30	0
sand 4	20.0	20.0	$1.0.10^{-6}$	$5.0.10^{-7}$	0.021	$3.0.10^{-3}$	$3.0.10^{-4}$	1	30	0

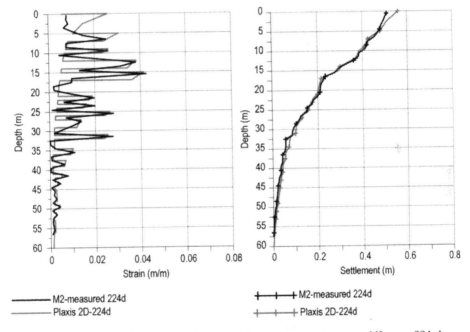

M2-measured 224d
Plaxis 2D-224d

M2-measured 224d
Plaxis 2D-224d

FIG. 9 *Measured and calculated (FEM) deformations in micrometer M2 at t = 224 days*

by reducing linearly the pressure between the hydrostatic value at a depth of 15 m and a reduction of 180 kPa at the gravel layer, 60 m deep. This is indicated in Fig. 10.

Once drained conditions are met a new distribution of OCR values, also shown in Fig. 10, is calculated. The elastoplastic model predicts a stiffer soil response when OCR increases and this is illustrated also in Fig. 9.

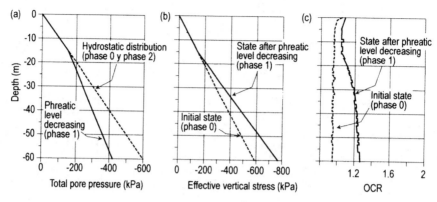

FIG. 10 *Past changes in water pressures and resulting OCR ratio. a) Assumed change in the theoretical hydrostatic profile; b) Associated distribution of vertical effective stresses at steady state; c) Calculated profiles of OCR*

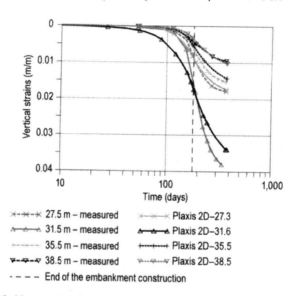

FIG. 11 *Measured and calculated vertical deformations at several depths*

The actual sequence of loading and consolidation periods (Fig. 3) was introduced in the model. Their response is compared in Fig. 12 with the micrometer – measured vertical displacements at several depths. The agreement is reasonably good.

The model is also capable of predicting unloading strains with a good approximation (Fig. 12).

3.2 One-dimensional model

The real preloading involves a very large area and therefore a simpler 1D model may prove accurate in practice at a reduced computational cost. In the model developed, only two materials are considered: clay (c) and sand (s).

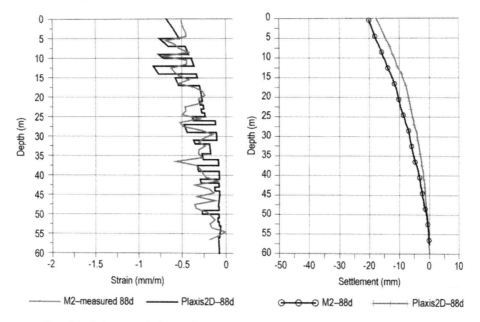

FIG. 12 *Calculated (FE model) and measured strains during unloading at t = 88 days after the start of unloading*

In both of them the effect of preconsolidation on primary and secondary compression rates is introduced.

The model is a superposition of 1D consolidation clay layers bounded by free draining sand layers. It may handle the following features:

- Variation of total stress with time. An approximate procedure to include 3D effects may be included by adjusting the total stress by means of elastic solutions.

- Available initial state of preconsolidation with depth.

- A generalized secondary consolidation model which includes overconsolidation and stress effects.

- Secondary effects on sand layers.

Calculation of excess pore water pressures:

For a total stress varying in time $\sigma(t)$, the excess pore pressure inside a consolidating layer is given by,

$$u(t,z) = \sum_{n=1,3,5...} 2\left(\frac{2}{n\pi}\right)^3 \sin\left(\frac{n\pi}{2}\frac{z}{H}\right)$$

$$\times \sum_i \left(\frac{d\sigma}{dt}\right)_i \frac{H^2}{c_v}\left[\exp\left(\left(\frac{n\pi}{2}\right)^2 (t_{ifin}-t)\frac{c_v}{H^2}\right) - \exp\left(\left(\frac{n\pi}{2}\right)^2 (t_{iini}-t)\frac{c_v}{H^2}\right)\right] \quad (1)$$

where H: half thickness of a double consolidating layer, i: the ith loading phase; $t_{i\,ini}$: initial time of the ith loading phase; $t_{i\,fin}$ final time of the ith loading phase; $(d\sigma/dt)$: a constant rate of load charge during the ith loading phase. Alonso and Krizek (1974) provide the theoretical background for the preceding equation.

Primary deformation:

The increment of primary deformation $\varepsilon_2^p - \varepsilon_1^p$ included by a change in vertical effective stress from σ_1' to σ_2' and a change in preconsolidation stress from σ_{c1}' to σ_{c2}' is given by

$$\varepsilon_2^p \varepsilon_1^p = \frac{1}{1 + e_0}[C_c(\log\sigma_{c2}' - \log\sigma_{c1}') + C_s(\log\sigma_{c2}' - \log\sigma_{c2}') - (\log\sigma_1' - \log\sigma_{c1}')] \quad (2)$$

Where e_0 is the initial void ratio.

Secondary deformations:

The increment of secondary deformations from time t_2 to time t_1 due to the elapsed time under stress σ_2' and an overconsolidation ratio OCR, at time t_1 is given by:

$$\varepsilon_2^s - \varepsilon_1^s = \begin{cases} \frac{\sigma_2' - \sigma_0'}{\sigma_{ref}'}(C_{\alpha min} + (C_{\alpha max} - C_{\alpha min}) \\ \quad \exp[-C_{\alpha dec}(OCR_1 - 1)]) \log\left(\frac{t_2 - t^*}{t_1 - t^*}\right) & \text{si } t_2 > 0.848\frac{H^2}{c_v} \\ 0 & \text{si } t_2 \leq 0.848\frac{H^2}{c_v} \end{cases} \quad (3)$$

where σ_0' is the effective vertical stress at the beginning of loading history ($t_0 = 0$); σ_{ref}' is a reference stress; $C_{\alpha max}$, $C_{\alpha min}$ are the maximum and minimum secondary coefficients for $\sigma' = \sigma_{ref}'$ and OCR = 1 and OCR = ∞ respectively; $C_{\alpha dec}$ is a transition parameter giving C_α as function of $C_{\alpha max}$ and $C_{\alpha min}$; and t^* is an auxiliary time variable used to smooth the solution.

Secondary settlements are computed for times in excess of time factor = 0.848 (90% degree of consolidation). A similar model is used for sand layers.

The model was implemented into an Excel sheet. The determination of the distribution of initial effective stress was made by lowering the water level 18 m in the lower gravels and allowing full consolidation to layers located between z = 15 m and z = 65 m.

Model parameters were derived by a similar trial-and-error procedure. Strains measured by micrometers (see one example in Fig. 13) could be precisely reproduced. The measured time history of surface settlements is compared with the one dimensional calculation in Fig. 14.

4   MODELLING FOR THE DESIGN STAGE

The two previous models were successfully calibrated with the help of the preloading test. They were capable of precise fittings. There was an interest in comparing the predictions of the two models in design calculations. In design

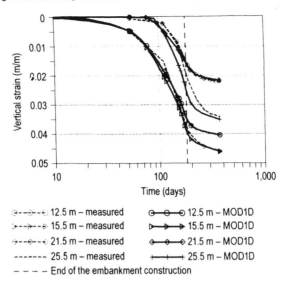

12.5 m – measured   12.5 m – MOD1D
15.5 m – measured   15.5 m – MOD1D
21.5 m – measured   21.5 m – MOD1D
25.5 m – measured   25.5 m – MOD1D
− − − − End of the embankment construction

FIG. 13 *Measured and calculated (1D model) strains in extensometer M-3*

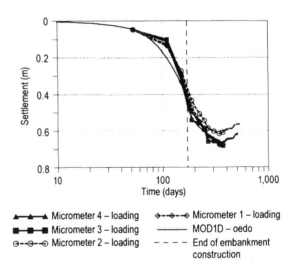

Micrometer 4 – loading   Micrometer 1 – loading
Micrometer 3 – loading   MOD1D – oedo
Micrometer 2 – loading   − − − − End of embankment construction

FIG. 14 *Measured and calculated (1D model) surface movements at the location of the four micrometers. Primary and secondary settlements are distinguished*

the decision to be made concerns the intensity and duration of preloading. In the cases reported as follows the intensity of preloading was induced by an embankment 8 m high.

The following design scenario was analyzed: three consolidation times were examined: 90, 180, and 360 days. After consolidation, the embankment was removed in 60 days. A pavement was then extended and the service loads were applied. Three service load intensities were considered (40, 60 and 80 kPa). The calculations were run for a long period (7.1 years) after the application of the service loads. In all the cases analyzed, the preconsolidation effect described previously (lowering the water table) was also introduced. The 2D case involved loading an infinite strip 110 m wide.

Consider the long term settlements induced by service loads. The loading history before the application of these loads involves the preloading embankments construction, a (variable) consolidation time, unloading and the application of a pavement loading. Then, the service load is applied in a ramp manner for 180 days.

The plot in Fig. 15 shows the settlements calculated (via FEM) during the last 30 days of service loading application and during the subsequent development of delayed settlements for a period of 7.1 years. The same plot, in the case of 1D calculation, is given in Fig. 16. Results are quite similar

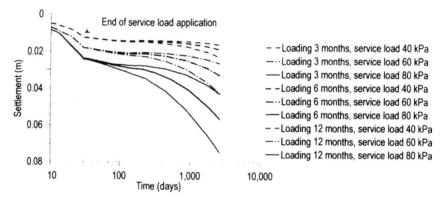

FIG. 15 *Calculated (FEM) evolution of long term settlements for varying preloading times and intensity of service loads*

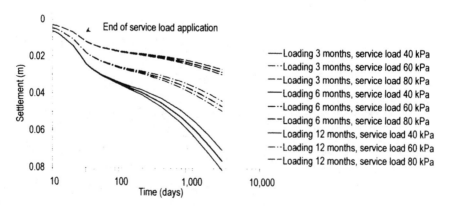

FIG. 16 *Calculated (1D) evolution of long term settlements for varying preloading times and intensity of service loads*

despite the large conceptual differences between the two models and this is an interesting outcome.

The preloading time helps to reduce the calculated long-term settlements especially in the 2D FEM calculations. However, this effect is not so marked in the 1D model.

5   CONCLUSIONS

This paper describes the methodology to investigate in detail the compressibility characteristics of a deep deltaic heterogeneous soil profile. Standard soil investigation techniques (borings, sample testing and penetration tests) are classical procedures to approach the settlement analysis. However, their accuracy cannot match the direct determination of soil deformability in 1 m intervals provided by precision extensometers associated with a preloading test.

The extensometers provide, in a sense, a compression test for each one of a large set of layers 1 m thick. Short term loading, unloading and long-term deformations can be accurately measured and interpreted. The technique may allow a direct interpretation of data, but if used in connection with soil models, it leads to accurate prediction tools which may later be applied to predicting soil behavior under other circumstances.

Two different approaches, a FEM technique using an elastoplastic creep model and a simpler one dimensional analysis, which may be refined to include accurate creep models, have been validated through the extensometer readings. They have been applied to the design stage of a large preloading scheme. Predicted long term settlements are quite similar in both approaches, a result which adds confidence to the modelling exercise performed. The main reason for this result probably lies in the advantages of validating the models through a comprehensive and well focused set of data directly related with the main problem under study.

There are, however, some differences when comparing the two procedures. It appears that the length of the preloading time and its effect in reducing later settlements is more significant in FE than in 1D calculations (compare Figs. 15 and 16). In addition, and despite the good validation of the two models, the 1D approximation tends to predict higher settlements than the FE model, especially for short consolidation times during the preloading stage. A more accurate evaluation of these aspects would require long term records of settlements under service loads.

ACKNOWLEDGEMENTS

The support provided by the Port Authority of Barcelona is greatly acknowledged by the authors.

REFERENCES

Alonso, E. E. and Krizek, R. J. (1974). Randomness of Settlement Rate under Stochastic Load. *Journal of the Engineering Mechanics Division*, ASCE, Vol. 100, EM6, pp. 1211-1226.

Alonso, E. E., Gens, A. and Lloret, A. (2000). Precompression design for secondary settlement reduction. *Géotechnique*, Vol. 50, No. 6, pp. 645-656.

Plaxis (2010). www.plaxis.nl.

FEM simulation of vacuum consolidation with CPVD for under-consolidated deposit

Jinchun Chai
Department of Civil Engineering, Saga University, Japan

Dennes T. Bergado
School of of Engineering and Technology, Asian Institute of Technology, Thailand

Takanori Hino
Institute of Lowland Technology, Saga University, Japan

KEYWORDS vacuum consolidation, finite element analysis, reclaimed land

ABSTRACT A method for simulating the vacuum consolidation with capped prefabricated vertical drain (CPVD) for under-consolidated deposits has been proposed. To establish a correct initial stress condition, the reclamation process is simulated by turning on the gravity force of fill material layer-by-layer and re-defining the drainage boundary condition after a new layer is placed. Further, the effect of installing CPVDs is modeled by changing the properties of the corresponding soil layer. The method was applied to a case history in Tokyo Bay, Japan. Comparing the simulated results with the measured data indicates that the method simulated the settlement curves reasonably well. As for the lateral displacement, the simulations overpredicted the observed values.

1 INTRODUCTION

Urban development in coastal area often involves land reclamation. Due to the short allowable time period, there are cases that the reclaimed lands with clayey fills are in an under-consolidated state before infrastructure constructions. For this kind of situations, preloading with prefabricated vertical drain (PVD) method is widely used to improve the ground condition. Recently, there are cases of using vacuum consolidation combined with Capped PVD (CPVD) to improve the reclaimed under-consolidated deposits (Nakaoka et al., 2005; Miyakoshi et al., 2007a).

Vacuum consolidation induces settlements as well as inward lateral displacements which are different from that of embankment loading. Chai et al. (2005) proposed a method for calculating vacuum pressure induced ground deformations and demonstrated successful applications to the case histories involving close to normally consolidated deposits. For under-consolidated deposit, the method under-predict the lateral displacement (Chai et al., 2008).

Finite element method (FEM) is a powerful tool for analyzing the boundary value problem in geotechnical engineering. FEM was used to analyze the ground deformation induced by vacuum consolidation of under-consolidated deposit, and by some kind of treatment, reasonable agreement between the field measured and predicted results has been reported (e.g. Mizuno et al., 2008). However, the analysis failed to simulate the initial under-consolidated state of the deposit, which is an important aspect when using elasto-plastic constitutive soil model.

In this paper, a method has been proposed to simulate the vacuum – CPVD consolidation induced ground deformation of under-consolidated deposits. The method was applied to a case history in Tokyo Bay, Japan (Miyakoshi et al., 2007a, b). The procedure of the proposed method is presented first followed the description of the case history and the comparison of the simulated results with the field measured data.

2   PROPOSED FEM PROCEDURE

It is proposed that a correct initial stress condition of an under-consolidated deposit for FEM analysis can only be established by starting simulation from the reclamation process. Since the vacuum consolidation normally combined with PVD improvement, the numerical method should be able to simulate the effect of PVD installation also. The proposed procedures are as follows:

(1) Simulating reclamation process. In the field, the reclamation is made by placing fill materials layer by layer. In FEM simulation, this process has to be followed closely by turning on the gravity force of the fill material layer by layer. In case of under-water reclamation, buoyancy unit weight should be applied. During this process, the drainage boundary condition of the problem changes with the application of each new layer of the fill material, i.e. the top surface is always a free drainage boundary, and FEM analysis should be able to simulate this phenomenon. Unfortunately, most commercial software can not treat this aspect properly. Another point is that the purpose of the simulation is to set-up a correct initial stress condition for a known ground geometry, and the deformation induced by the self-weight during the reclamation period has to be cleared up at the end of each step.

In simulating the reclamation process, when a new layer of fill is placed (turn on the gravity force), the initial effective stress is close to zero and it is difficult to use an elasto-plastic soil model. To overcome this problem, initially, a newly placed fill material is treated as a linear elastic material, and after CPVD installation it has been changed to an elasto-plastic soil model.

(2) Simulating PVD installation process. There are two methods for simulating the PVD improved deposit. One is using drainage elements to simulate the effect of PVD (Hird et al., 1992; Chai et al., 1995; Indraratna and Redana, 2000) and another is using an equivalent vertical hydraulic conductivity (k_{ev}) to model the effect of PVD (Chai et al., 2001). In case of using drainage element method, if not adopting re-meshing technique, all drainage elements have to be included into the mesh during the simulation of reclamation pro-

cess and it is troublesome. It is recommended to use the equivalent vertical hydraulic conductivity method (Chai et al., 2001). Effect of PVD installation is simulated by changing the material properties before and after the installation.

The above features have been incorporated into the finite element code M-CRISP and used for this study. M-CRISP is a modified form of the original code CRISP (Britto and Gunn, 1987).

3   BRIEF DESCRIPTION OF A CASE HISTORY IN TOKYO BAY, JAPAN

In Tokyo Bay new landfill site, Japan, the soil deposits consist of the original clayey soil at the sea bed with a thickness of about 30 m. In recent years, the dredged clayey soils were dumped in the site and up to October 2005, the thickness of the dredged material reached to about 12 m with a rate of about 3.5 m/year. At the beginning of 2006, the original clayey deposit and part of the dredged layer was in an under-consolidated state and the surface of the landfill was close to the sea level (Takeya et al., 2007). To increase the capacity of the landfill, there was a need to consolidate the soft clayey deposit to reduce its volume. It has been considered that vacuum consolidation combined with CPVDs, which will be referred as vacuum pressure-CPVD method (Chai et al., 2008), is a suitable method for this situation. The method applies vacuum pressure to each CPVD and uses a clayey surface layer as air-tightening layer instead of a air-tightening sheet. From January 2006 to January 2007, two test sections at Tokyo Bay, Japan, had been conducted using the vacuum pressure-CPVD method (Takeya et al., 2007; Miyakoshi et al., 2007a).

The plan-view of the two test sections is shown in Fig. 1 together with the locations of key field instrumentation points (modified from Takeya et al., 2007). Section-A had an area of 60 m × 60 m and Section-B of 61.2 m × 61.2 m. The soil profile at the test site at the end of 2005 is shown in Fig. 2

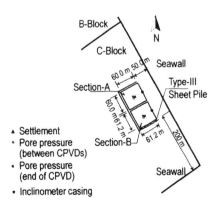

FIG. 1 *Plan-view of the test areas*

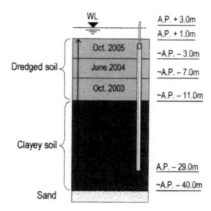

FIG. 2 *Soil profiles at the test site*

(modified from Miyakoshi et al., 2007a). The total unit weight (γ_t), compression index (C_c) and maximum consolidation pressure (p_c) before the CPVDs installation are shown in Fig. 3 (data from Miyakoshi et al., 2007a). It can be seen that the dredged (reclaimed) layer was in a state of close to normally consolidation but the original clayey deposit was in an under-consolidated state. In the original deposit, the clay (grain size less than $5\,\mu$m) content is more than 50% and for the dredged layer, the sum of sand and silt contents is slightly more than 50% as shown in Fig. 4 (data from Miyakoshi et al., 2007a).

The CPVDs used at this site had a cross-section of 150 mm by 3 mm. At Section-A, the CPVDs had a spacing of 2.0 m and 1.8 m for Section-B with a square pattern. For both the sections, the CPVDs were installed to 30 m depth from the ground surface, and the air-tightening surface clayey soil layer had a thickness of about 1.0 m. The field installation started at the beginning of January 2006 in Section-A, and at the beginning of February 2006 in Section-B, and for both the sections, the duration of installation was about one month.

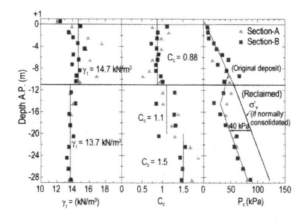

FIG. 3 γ_t, C_c and P_c of the deposit

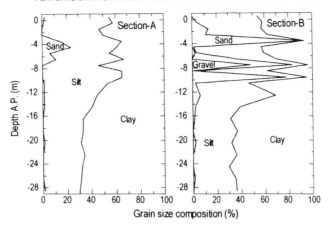

FIG. 4 *Grain size compositions*

The duration of after the CPVD installation and before application of vacuum pressure was about 5 and 4 months for Section-A and B, respectively. Considering the half period of the installation as consolidation time, the partial self-weight consolidation period after the CPVD installation and before vacuum pressure application was about 165 and 135 days for Section-A and B, respectively. From June 30, 2006, the vacuum pressure with a value of 80 to 90 kPa at the vacuum pump location was applied and lasted for 204 days. Surface and subsurface settlement gages, excess pore water pressure (vacuum) gages at the vacuum pump and at the ends of two CPVDs, as well as inclinometer casings were installed to monitor the ground response. At the earlier stages of vacuum consolidation, the vacuum pressure at the ends of CPVDs was about −50 kPa and at the later stages, it reached about −70 to −75 kPa.

Takeya et al. (2007) reported that for both the original clayey deposit and the dredged layer, the vertical coefficient of consolidation (C_v) was about 0.012 m²/day. However, from the information in Figs. 3 and 4 (Pc values and the grain size distribution), it can be argued that the dredged layer may have a higher C_v value. For calculating the consolidation settlement, $C_v = 0.012$ m²/day for the original deposit and 0.024 m²/day for the dredged layer were used by Chai et al. (2010). Further, assuming that the coefficient of consolidation in the horizontal direction is twice of the corresponding value in the vertical direction and the discharge capacity of the CPVD, $q_w = 500$ m³/year, a ratio between the horizontal hydraulic conductivity (k_h) to that in smear zone (k_s) $k_h/k_s = 2$ was back evaluated.

4   FEM SIMULATION

4.1 Modeling vacuum pressure – CPVD method

For the vacuum pressure – CPVD method, vacuum pressure is applied to each CPVD at the cap location and the average vacuum pressure at the level of the cap is smaller than the value applied to each CPVD. To approximately consider this factor, vacuum pressure was applied to every other node in horizontal direction at the level of the cap of CPVD. Referring the field measured data, the applied vacuum pressure was -70 kPa.

To considering the possible occurrence of tension cracks around the edge of the vacuum consolidation area, six (6) nodes permeable joint elements were used around the edge of the improved area.

4.2 Model parameters

In FEM analysis, the original deposit was represented by Modified Cam clay model (Roscoe and Burland, 1968). For the reclaimed layer, during filling process it was simulated as an elastic material with a Young's modulus $E = 1000$ kPa and Poisson's ratio $\mu = 0.4$, and after CPVD installation it was changed to modified Cam clay model except the 1 m thick surface layer, which was treated as an elastic material but μ was reduced to 0.3. The values of model parameters are listed in Table 1. The parameters related to CPVD consolidation are given in Table 2.

TABLE 1 MODEL PARAMETERS

Depth (m)	Soil layer	γ_t* (kN/m³)	e_0	λ	κ	M	E (kPa)	μ	k_v	k_h (10⁻³ m/day)
0 - 1	Reclaimed	14.7					1,000	0.3	22.6	45.3
0 - 12	Reclaimed	14.7	2.41	0.38	0.038	1.2	—	0.3	22.6	45.3
12 - 15	Clay 1	13.7	3.28	0.48 / 0.10#	0.048 / 0.01#	1.2	—	0.3	4.23	8.46
15 - 21	Clay 2	13.7	3.28	0.48	0.048	1.2	—	0.3	4.23	8.46
21 - 30	Clay 3	13.7	3.28	0.65	0.065	1.2	—	0.3	1.65	3.33
30 - 41	Clay 4	13.7	3.28	0.65	0.065	1.2	—	0.3	0.78	1.56

Note: *γ_t, unit weight; e_0, initial void ratio; λ, slope of virgin consolidation line in $e - \ln p'$ plot (e is void ratio and p' is effective mean stress); κ, slope of rebound line in $e - \ln p'$ plot; M, slope of critical state line in $p' - q$ plot (q is deviator stress); E, Young's modulus; μ, Poisson's ratio; k_h and k_v, horizontal and vertical hydraulic conductivities. # The values are assumed for Case-A2 and B2.

TABLE 2 PARAMETERS RELATED TO CPVD CONSOLIDATION

Parameter	Symbol	Unit	Value A	Value B
Unit cell diameter	D_e	m	2.26	2.03
Drain diameter	d_w	mm	75	
Smear zone diameter	d_s	m	0.3	
Discharge capacity	q_w	m³/day	1.37	
Hydraulic conductivity ratio	k_h/k_s#	—	8	

#k_s is the hydraulic conductivity of the smear zone.

The values of γ_t, e_0 and λ were determined based on the test data reported by Miyakoshi et al. (2007a). The values of κ, M and μ were assumed empirically. The values of hydraulic conductivities (k_h and k_v) in Table 1 are initial ones and during consolidation process, they were varied with void ratio of soil following Taylor's (1948) equation:

$$k = k_0 \cdot 10^{-(e_0-e)/C_k} \tag{1}$$

where k_0 is the initial hydraulic conductivity, k is the hydraulic conductivity corresponding to e, and C_k is a constant and equals $0.4e_0$ in this study.

The initial k_v values were evaluated from C_v as well as coefficient of constrained modulus (m_v).

$$k_v = c_v \gamma_w m_v \tag{2}$$

where γ_w is the unit weight of water. In evaluating the value of m_v, for the original deposit, the effective vertical stress considered was that induced by the buoyancy gravity force of the original deposit (without dredged fill material condition), and for the dredged fill, the stress induced by the buoyancy gravity force of 1 m thick fill. However, if using the back-evaluated C_v values from Chai et al. (2010), to obtain a stress state close to that shown in Fig. 3 before the start of the project, a filling rate of 1 m/year was resulted, which is much less than the reported about 3.5 m/year (Takeya et al., 2007). The initial k values listed in Table 1 were deduced using the C_v values of three (3) times of the values reported by Chai et al. (2010). The adopted C_v value is much larger than the reported laboratory value (Takeya et al., 2007). It is generally accepted that the laboratory consolidation test under-evaluates the field coefficient of consolidation because it can not consider the effect of stratification of a field deposit (Chai and Miura, 1999). However, for this case the degree of consolidation is not only influenced by C_v value of the deposit but also the parameters related to CPVD consolidation. If using the initial k values in Table 1, the simulated degree of consolidation during vacuum consolidation will be much faster than the field data. To overcome this dilemma, the parameters related to CPVD consolidation were re-evaluated. Keeping $q_w = 500\,\mathrm{m^3/year}$ and a $k_h/k_s = 8$ was newly estimated (a value of 2 by Chai et al., 2010). Note, with the k_v values in Table 1, $E = 1,000\,\mathrm{kPa}$, $\mu = 0.4$, a C_v value of about $5\,\mathrm{m^2/year}$ can be resulted for the dredged layer. This emphasizes that during the reclamation process, the fill material might have a larger void ratio and therefore a higher hydraulic conductivity.

A hyperbolic model was used to simulate the behavior of the joint elements. The tangential shear stiffness k_t was calculated by the following equation:

$$k_t = \left(1 - \frac{R_f \tau}{c_a + \sigma_n \tan \delta}\right)^2 k_1 \gamma_w \left(\frac{\sigma_n}{p_a}\right)^n \tag{3}$$

where c_a is cohesion, δ is interface friction angle, τ is shear stress, σ_n is normal stress, p_a is the atmosphere pressure, R_f is failure ratio (= 1 here), and k_1 and n (= 1 here) are constants. The assumed values of c_a, δ and k_1 are listed in Table 3. The joint was allowed to open if the normal stress of the joint becomes negative.

TABLE 3   PARAMETERS FOR JOINT ELEMENTS

Depth (m)	c_a (kPa)	δ (°)	k_1
0~6	10	30	200
6~12	15	30	300
12~21	20	30	300
21~30	40	30	1,000

4.3 Cases simulated

Two simulations were conducted for each section. Case-A1 and B1 used the parameters in Table 1 and 2, respectively. However, since the analyses did not simulate the lateral displacement well, Case-A2 and B2 used a smaller λ value (see the note # below Table 1), the value in the shadowed zone of Table 1 for the $0 - 3\,\mathrm{m}$ depth of the original clayey deposit (12 to 15 m depth from the ground surface before the vacuum consolidation).

5   RESULTS

5.1 Effective vertical stress

The simulated effective vertical stresses (σ'_{v0}) in the ground before the CPVD installation are compared in Fig. 5 (measured data from Miyakoshi et al., 2007a). It can be seen that Case-A1 and B1 simulated well the field condition. However, Case-A2 and B2 resulted in higher σ'_{v0} values in the surface layer of the original deposit. This comparison indicates that the proposed method is effective for establishing an under-consolidated initial stress condition.

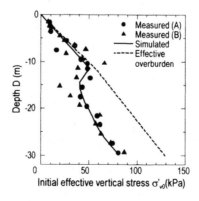

FIG. 5 *Effective vertical stresses*

5.2 Vacuum pressure in the ground

In the field, only the vacuum pressures applied to the top of CPVD and at the end of two instrumented CPVDs were measured. Using the equivalent vertical hydraulic conductivity, at a given depth, the simulated vacuum pressure represents average values within the improved zone of a CPVD and cannot be directly compared with the measured data within the CPVD. As a reference, Figs. 6 (a) and (b) and 7 (a) and (b) show the measured values at the end of the instrumented CPVDs and the simulated values in the ground for Section-A and B, respectively. The measured data are from Miyakoshi et al. (2007a). The simulated final vacuum pressure in the ground is about −50 kPa and about 70% of the applied value at the cap location of the CPVDs. As a general tendency, the equivalent vertical hydraulic conductivity method adopted tends to under-predict the vacuum pressure at the bottom of the ground.

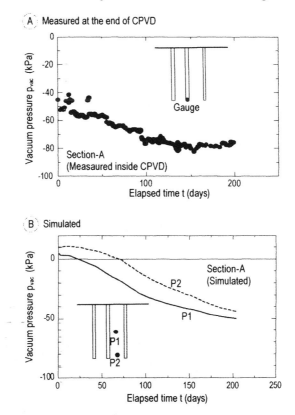

FIG. 6 *Vacuum pressure inside CPVD and in the ground (Section-A)*

5.3 Settlement curves

The comparison of settlement curves are depicted in Figs. 8 and 9 for Section-A and B, respectively (measured data from Miyakoshi et al., 2007a). Generally analyses (Case-A1 and B1) fairly simulated the measured total settlement

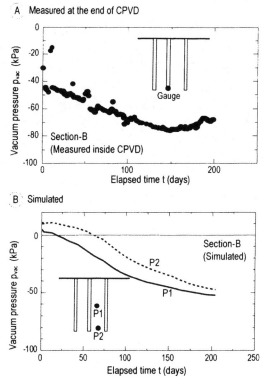

FIG. 7 *Vacuum pressure inside CPVD and in the ground (Section-B)*

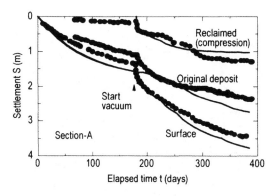

FIG. 8 *Settlements of Section-A*

curves. For after CPVDs installation and before application of vacuum pressure period, for both the sections, the simulations under-predicted the compression of the reclaimed layer and overpredicted the settlement of the original clayey deposit. During the vacuum consolidation period, for Section-B, the simulation under-predicted the compression of the reclaimed layer also.

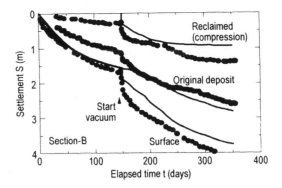

FIG. 9 *Settlements of Section-B*

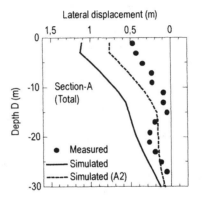

FIG. 10 *Lateral displacements of Section-A*

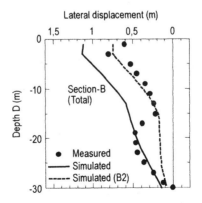

FIG. 11 *Lateral displacements of Section-B*

5.4 Lateral displacement

The lateral displacements at the edge of the vacuum consolidation area (see Fig. 1 for locations) are compared in Figs. 10 and 11 for Section-A and B, respectively (measured data from Miyakoshi et al., 2007a). The agreement between the simulated and measured values is poor. The measurement shows that the lateral displacement at around the surface of the original deposit is quite smaller, but the simulated results (Case-A1 and B1) do not show this tendency. Although the site investigation does not indicate a stronger surface layer of the original deposit, Case-A2 and B2 were conducted by assuming a smaller λ value (0.1) for $0 - 3$ m depth original deposit, and the simulated lateral displacements (dashed lines) become closer to the measured ones. However, Case-A2 and B2 under-predicted the settlements of about 0.5 m for both the sections.

6 CONCLUSIONS

A method for simulating the vacuum consolidation combined with capped prefabricated vertical drain (CPVD) for under-consolidated deposits has been proposed. To establish a correct initial stress condition, the reclamation process is simulated by turning on the gravity force of fill material layer-by-layer and re-defining drainage boundary condition after a new layer is placed. In this process, the displacement is cleared up at the end of each step. Further, the effect of installing CPVD is modeled by changing the properties of the corresponding soil layer.

The method was applied to a case history in Tokyo Bay, Japan. Comparing the simulated results with the measured data indicates:

(a) To obtain a correct initial effective vertical stress in the deposit, during the simulation of the reclamation process, a higher vertical coefficient of consolidation (C_v) for the reclaimed layer needs to be used. For the case studied, the value adopted is about 50 times of that used to simulate the vacuum consolidation.

(b) The proposed method simulated the settlement curves reasonably well. As for the lateral displacement, the simulations overpredicted the observed values. Further research may be needed to investigate this aspect.

REFERENCES

Britto, A. M. and Gunn, M. J. (1987). *Critical state soil mechnics via finite elements.* Ellis Horwood Limited, p. 486.

Chai, J.-C., Miura, N., Sakajo, S., and Bergado, D. T. (1995). Behavior of vertical drain improved subsoil under embankment loading. *Soils and Found.*, 35(4), 49-61.

Chai, J.-C. and Miura, N. (1999). Investigation on some factors affecting vertical drain behavior. *Journal of Geotechnical and Geoenvironmental Engineering*, ASCE 125(3), 216-226.

Chai, J.-C., Shen, S.-L., Miura, N. and Bergado, D. T. (2001). Simple method of modeling PVD improved subsoil. *Journal of Geotechnical and Geoenvironmental Engineering*, ASCE, 127(11), 965-972.

Chai, J.-C., Carter, J. P. and Hayahsi, S. (2005). Ground deformation induced by vacuum consolidation. *Journal of Geotechnical and Geoenvironmental Engineering*, ASCE, 131(12), 1552-1561.

Chai, J.-C., Miura, N. and Bergado, D. T. (2008). Preloading clayey deposit by vacuum pressure with cap-drain: Analyses versus performance. *Geotextiles and Geomembranes*, 26(3), 220-230.

Chai, J.-C., Hino, T. and Miura, N. (2010). Deformation calculation of under-consolidated deposit induced by vacuum consolidation. Accepted for *Proceedings of the 9th Intl. Conference on Geosynthetics*, Brazil.

Hird, C. C., Pyrah, I. C. and Russell, D. (1992). Finite element modelling of vertical drains beneath embankments on soft ground. *Geotechnique*, 42(3), 499-511.

Indraratna, B. and Redana, I. W. (2000). Numerical modeling of vertical drains with smear and well resistance installed in soft clay. *Canadian Geotechnical Journal*, 37, 132-145.

Nakaoka, J., Fujiki, Y., Yoneya H., Shinsha, H. (2005). Application of vacuum consolidation for improvement of soft ground reclaimed with dredged clayey soils: effect of the improvement and the strength of the ground. *Proceedings of 40th Annual Meeting*, Japanese Geotechnical Society, Hakodate, Japan, 1055-1056 (in Japanese).

Miyakoshi, K., Shinsha, H. and Nakagawa, D., (2007a). Vacuum consolidation field test for volume reduction scheme of soft clayey ground (Part-2) – ground characteristic and measured results. *Proceedings of 42th Annual Meeting*, Japanese Geotechnical Society, 919-920 (in Japanese).

Miyakoshi, K., Takeya, K., Otsuki, Y., Nozue, Y., Kosaka, H., Kumagai, M., Oowada, T. and Yamashita, T. (2007b). The application of the vacuum compaction drain method to prolong the life of an offshore disposal field. *Nippon Koei Technical Forum*, No. 16, 9 – 19 (in Japanese).

Mizuno, K., Tsuchida T. and Shisha, H. (2008). Ground deformation of dredging clay reclaimed land by vacuum consolidation method and its finite element analysis. *Geotechnical Journal,* Japanese Geotechnical Society, 3(1), 95-108 (in Japanese).

Roscoe, K. H. and Burland, J. B. (1968). On the generalized stress-strain behaviour of 'wet' clay. *Engineering Plasticity* (Edited by J. Heyman and F. A. Leckie), Cambridge University Press, 535-609.

Takeya, K., Nagatsu, T., Yamashita, T. (2007). Vacuum consolidation field test for volumn reduction scheme of soft clayey ground (Part-1) – field test condition and construction method. *Proceedings of 42th Annual Meeting*, Japanese Geotechnical Society, 917-918 (in Japanese).

Taylor, D. W. (1948). *Fundamentals of Soil Mechanics*, John Wiley & Sons Inc. New York.

Improvement of ultra soft soil for the reclamation of a slurry pond

Jian Chu
Centre for Infrastructure Systems, Nanyang Technological University, Singapore

Mint Win Bo
Director (Geo-services), DST Consulting Engineers Inc., Thunder Bay, Ontario, Canada

Arul Arulrajah
Swinburne University of Technology, Melbourne, Australia

KEYWORDS land reclamation, prefabricated vertical drain, soft soil, soil improvement

ABSTRACT A case study for the improvement of an ultra soft soil as part of a land reclamation project in Singapore is presented in this paper. The ultra soft soil was in a slurry pond, which covered an area of 180 ha. The soil was recently deposited in the pond as slurry. It was high in water content and had almost zero shear strength. The reclamation was carried out by firstly spreading sand fill in thin layers of 20 cm thick using a specially designed sand spreader. As this method was not successful in some areas, geotextile sheets were used to cover an area of $630,000\,m^2$ before more sand fill was placed. After the completion of fill placement, fill surcharge and prefabricated vertical drains (PVDs) were used to improve and accelerate the consolidation of the slurry. As the performance of PVDs would deteriorate after they had undergone large deformation, the PVDs were installed in two stages. In the first stage, PVDs were inserted with a square grid spacing of 2.0 m. After nearly 1.5 m of settlement had taken place, the second stage of PVDs with the same spacing was installed at the centre of the square grid of the PVDs installed in the first stage. After nearly 4 years of consolidation, the top of the slurry had settled more than 3.5 m. The undrained shear strength of the soil had also increased substantially.

1 INTRODUCTION

As part of a 2,000 ha reclamation project at Changi East, Singapore, a 180 ha slurry pond needed to be reclaimed. A picture of the slurry pond, taken before reclamation, is shown in Fig. 1. The slurry pond was trapezoidal in shape, approximately 2,000 m long and 750-1,050 m wide. It was created by dredging the original seabed to an elevation of −22 mCD (Admiralty chart datum, where mean sea level is at +1.6 mCD) between 1975 and 1978 as a borrow pit. A containment sand bund was constructed around this borrow pit in 1986 to the crest level of about +5 mCD. Subsequently, silt and clay washings from other sand quarrying activities in the eastern part of Singapore were transported through pipelines with water and discharged into this contained area to form a pond. Therefore, the slurry inside the pond consisted mainly of clay and silt. The water level in the pond was at +3 mCD. During the subsequent years, suspension in the slurry had settled in the pond, and the majority of the slurry was undergoing self-weight consolidation.

FIG. 1 *The slurry pond before reclamation*

2   SITE CONDITIONS AND PROPERTIES OF THE SLURRY

A cross-section of the soil profile in the slurry pond along Chainage X2600 is shown in Fig. 2. The elevation of the top of the slurry varied from −1 to −5 mCD. There was no clear boundary between water and slurry as the transition was rather gradual. The top of the slurry was taken as the elevation at which the density was greater than 1.1 Mg/m³. The bottom of the slurry was in the range of 0 to −22 mCD. The thickness of the slurry ranged between 3 to 20 m.

The slurry material consisted mainly of high plasticity clay. The liquid limit, plastic limit, and moisture content distribution with depth are shown in Fig. 3a. The water content of the slurry varied from 60 to 300%, but was mainly in the range of 120 to 180%. The bulk density was low and the values ranged from 1.2 to 1.4 Mg/m³. The slurry was still undergoing self-weight consolidation. The undrained shear strength profile of the slurry as determined by field vane shear tests is shown in Fig. 3b. The majority of the data points in Fig. 3b show an increasing trend with depth at a rate of 0.2 kPa per meter depth. Even though, the values were less than 8 kPa for soils at 20 m deep. The grain size distribution of the slurry is shown in Fig. 4. It can be seen that the upper bound of D50 was 0.024 mm, but mostly in the range smaller than 0.001,

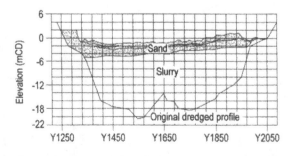

FIG. 2 *Cross-section of the slurry pond along Chainage X2600 and the sand fill placed before failure*

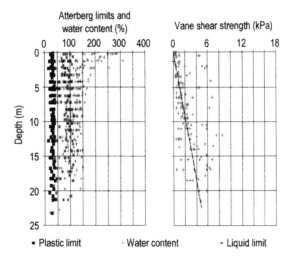

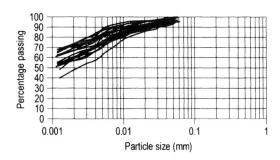

FIG. 3 *Variations of basic properties of the slurry with depth (a) Liquid limit, plastic limit and water content; and (b) Vane shear strength*

FIG. 4 *Grain size distributions of the slurry*

and D85 was in the range of 0.004 to 0.02 mm. The fines content was in the range of 70 to 93%.

3 RECLAMATION WORKS

As the top surface of the slurry had little strength, a direct hydraulic placement of sand fill on top of the slurry would cause the slurry to be displaced by the sand fill placed on top. As such, the sand fill had to be spread in thin layers using a specially designed sand spreader, as shown in Fig. 5. The sand fill with a high water to sand ratio was pumped into the spreader using a suction dredger. The sand spreading was carried out by moving the spreader repeatedly from left to right along the X-direction (the longer dimension). Small lifts of 20 cm were used in the first phase of spreading to ensure the stability of the fill. A waiting time was given between the placement of each 20 cm lift to allow the slurry to gain strength under the small surcharge of the sand fill. A hydrographic survey carried out along Chainage X2600, when the sand layer

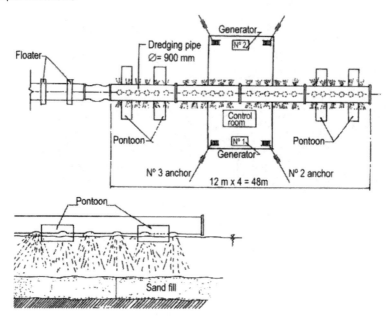

FIG. 5 *Sand spreading system used for the slurry pond*

reached 0 mCD is shown in Fig. 2. When the fill reached an elevation between 0 to +2 mCD, a failure in the form of slurry bursting occurred at the location as shown in Fig. 6.

The failure was attributed to uneven settlements and uneven spreading of the sand fill. It can be seen from Fig. 2, that the thickness of the slurry layer varied considerably across both the X and Y directions. The amount of compressibility of the slurry layer was thus different across both the X and Y directions even when the applied load was the same. The pore water pressure dissipation rates would also be different. The profile of the sand fill before failure along X2600 is shown in Fig. 2. Although the sand fill was spread carefully, a 1 m or so difference in the sand layer thickness could have been resulted. The difference in loads coupled with different pore pressure dissipation rates

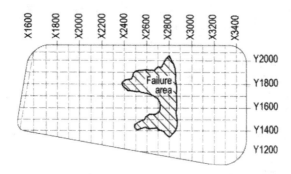

FIG. 6 *Location of the failure area where the bursting of slurry had occurred*

in soil led to the upheaval of slurry at some locations. This in turn displaced the sand above the upheaved slurry to two sides and subsequently caused the overburden pressure provided by the sand to be reduced. When the pore pressure became higher than the overburden, a mud burst occurred.

As a remedial measure, a geotextile fabric was used to cover the failure area. Geotextile has also been used for other similar projects by Broms (1987) and Terashi and Katagiri (2005). Before this, the burst mud was partially removed using a high capacity submersible mud pump. Based on analysis (Na et al., 1998; Chu et al., 2009), a tensile strength of 150 kN/m in both warp and weft directions was determined. Two types of woven geotextile were used. The first type, HS150/150, had a tensile strength of 150 kN/m in both directions. It was placed in a single layer to cover an area of 700 m × 300 m, as shown in Fig. 7 as area 2 in the left. Due to the shortage of supply of this type of geotextile, a second type, HS150/50, with tensile strengths of 100 kN/m and 50 kN/m in the warp and weft directions respectively was also used. By overlaying two layers with different warp and weft directions, a combined 150 kN/m tensile strength was achieved. The second type of geotextile was placed in two layers over an area of 700 m × 600 m, shown in Fig. 7 as area 3 in the right. Therefore, the total area covered by the geotextile was 700 m × 900 m or 630, 000 m^2.

The installation of the geotextile sheets was carried out in the following sequence: the supplied geotextile was in rolls of 5 m wide and 90 m long. They were sewn together using portable sewing machines to form a 700 m × 600 m or 700 m × 900 m geotextile sheet. When the geotextile was sewn together to form one piece, it was extremely heavy and required huge magnitude of forces to be pulled during installation. The total weight of the geotextile used was 320 tonnes. To overcome this problem, the geotextile sheet was folded into 10 m wide zigzag strips. This made the unfolding of the geotextile sheet much easier. The geotextile sheet was anchored at one end by placing sand on top of it. The geotextile was then folded back to cover the sand, which would be covered again by sand fill placed subsequently. The other end of the

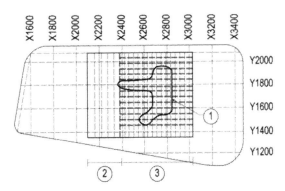

FIG. 7 *Layout of the geotextile sheets used for the failure area of the slurry pond*

geotextile sheet was connected to pipelines floating on water. The pipelines were pulled slowly toward the other side of the pond using 7 numbers of bulldozers, as shown in Fig. 8. Plastic buoys were also used for the inner area of the geotextile sheet to keep it floating on top of the water. When the geotextile sheet reached the other side of the pond, it was fixed in place by placing sand fill on top in a similar manner as it was for the other end. For the area where double layers of geotextile were used, the second layer was placed in the same way after the first layer had been installed. After the geotextile sheet was properly anchored at both ends, the plastic buoys were removed to allow the geotextile sheet to sink. At the interface of the two areas where the two different geotextiles were used, an overlapping length of 50 m was provided. A 50 cm thick sand layer was placed on the overlapping area for anchoring.

FIG. 8 *Installation of the geotextile sheet in the slurry pond*

After the placement of the geotextile, the sand fill spreading resumed using the same sand spreader until the ground elevation of +4 mCD was reached. In order to spread sand up to +4 mCD level using the spreader, the containment bund surround the slurry pond was elevated to +6 mCD and the water level to +5.5 mCD by pumping in sea water. It was still carried out in stages with the thickness of each layer controlled within 50 cm.

4   SOIL IMPROVEMENT WORKS

Before the PVDs installation at the +4 mCD level, the water level in the slurry pond was lowered to +3 mCD. Colbond drains CX1000 were installed with 2 m × 2 m square spacing as a first stage. The procedure discussed in Chu et al. (2004) was followed in selecting the type of drain and drain spacing. The surcharge at the first stage was placed to +6 mCD. The settlement of the fill was monitored using an instrumentation scheme as detailed by Bo et al. (2005). After approximately 1.5 m of settlement had taken place, a second stage of vertical drain were installed with the same square grid of 2.0 m in the centre of the square grid of the PVDs installed in the first round. The combined effective drain spacing is 1.4 m × 1.4 m. It was necessary to install the PVDs in two stages as the large ground settlement would cause the drains to buckle and thus affect the performance of the vertical drain. The deterioration in the performance of the vertical drain installed in the first stage was indicated by the fact that the rate of settlement had reduced to 20 cm/mth, although the pore water pressure remained very high in the soil. The installation of PVD itself assisted in the dissipation of the pore water pressure, as mud was seen to come out through the annulus of the mandrel (Chu et al., 2006). This is another advantage of the two stage PVD installation method. The final surcharge level of +9 mCD was placed after the installation of the second round of vertical drain.

The settlement and pore water pressure dissipation within the first 16 months are shown in Fig. 9. The positions of the settlement and pore pressure measuring devices are shown in Fig. 10. It can be seen from Fig. 9 that within the first 16 months, there was 2.7 m of ground settlement, but the pore pressure dissipations were slow.

Using Fig. 9, the pore pressure distribution in 16 months is plotted in Fig. 11. Based on Fig. 11, the average degree of consolidation can be estimated as 42%. Using this value and taking the thickness of the clay slurry layer as 7.8 m and double drained, the c_h can be back calculated as 0.27 m^2/yr, when c_v is assumed to be 0.1 m^2/yr. Applying Asaoka's method to the settlement data in Fig. 9b, the average degree of consolidation calculated is 91%, which is apparently too high. Therefore, the Asaoka's method is not applicable in this case where large deformation of sediment was encountered.

The much delayed dissipation of pore pressure in the slurry is typical of consolidation of ultra-soft soil as the slurry may have to undergo sedimentation and self-weight consolidation stage before it can be transformed into "soil". The Mandel-Cryer effect and the non-uniform consolidation of soil around PVDs could also be the other reasons.

Due the slow dissipation in pore pressure, it took several years to complete the primary consolidation despite the use of PVDs at an effective spacing of 1.4 m × 1.4 m by combining both passes of PVD installation. The complete loading history and surface settlement curves are shown in Fig. 12. As can be seen in Fig. 12a, one more surcharge increment of 5.5 m was applied at 2.3 years time. The additional settlement induced by this surcharge increment was only about 0.5 m, which was much smaller than the nearly 3 m settlement developed in the preceding stage. This indicates that the soil had been much improved over the past 2.3 years. The total settlement in 4 years time after preloading was 3.5 m and the degree of consolidation achieved was 85%. The undrained shear strength of the soil also increased considerably as reported by Chu et al. (2009).

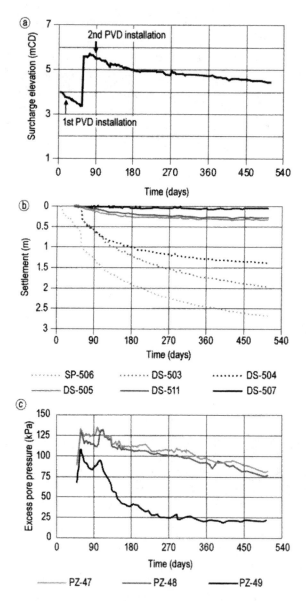

FIG. 9 *Monitoring data measured during the reclamation of slurry pond: (a) surcharge variation versus time; (b) settlement versus time; (c) excess pore pressure versus time*

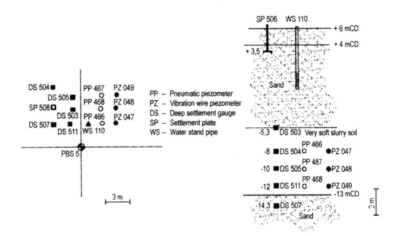

FIG. 10 *Arrangement of Instrumentation (a) plan view and (b) elevation view*

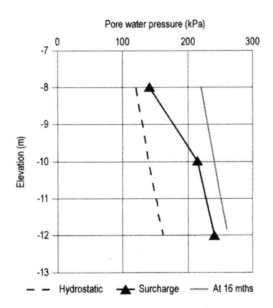

FIG. 11 *Pore water pressure profile 16 months after surcharge*

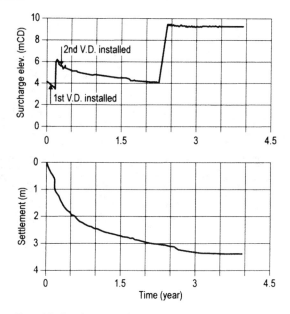

FIG. 12 *Surcharge and settlement versus time curves*

5 CONCLUSIONS

A case study for the reclamation of a slurry pond in Singapore is presented in this paper. The study shows that reclamation over ultra-soft slurry can be carried out. Although it is possible to place sand fill directly onto the slurry in thin layers, this may cause the slurry to burst, as shown by the case presented in this paper. In this case, geotextile sheets might have to be used as a separation layer before sand fill can be placed on slurry soil. PVDs and surcharge loads can be used for subsequent treatment of the slurry soil capped by fill surcharge.

REFERENCES

Bo, M. W., Chu, J. and Choa, V. (2005). The Changi East reclamation project in Singapore, *Ground Improvement – Case Histories*, Eds. B. Indraratna and J. Chu, Elsevier, 247-276.

Broms, B. B. (1987). Stabilization of very soft clay using geofabric. *Geotextiles and Geomembranes*. Vol. 5, 17-28.

Chu, J., Bo, M. W. and Choa, V. (2004). Practical consideration for using vertical drains in soil improvement projects, *Geotextiles and Geomembranes*, Vol. 22, No. 3, 101-117.

Chu, J., Bo, M. W. and Choa, V. (2006). Improvement of ultra-soft soil using prefabricated vertical drains, *Geotextiles and Geomembranes*, Vol. 24, No. 6, 339-348.

Chu, J., Bo, M. W. and Arulrajah, A. (2009). Reclamation of a slurry pond in Singapore, *Geotechnical Engineering, Proc ICE*, Vol. 162, GE1, 13-20.

Na, Y. M., Choa, V. Bo, M.W., and Arulrajah, A. (1998). Use of geosynthetics for reclamation on slurry like soil foundation, *Problematic Soils*, Yonagisawa et al. (eds.), Balkema, 767-771.

Terashi, M. and Katagiri, M. (2005). Key issues in the application of vertical drains to a sea reclamation by extremely soft clay slurry, *Ground Improvement – Case Histories*, Eds. B. Indraratna and J. Chu, Elsevier, 145-158.

Reinforced test embankments on Florianopolis very soft clay

Almeida, M. S. S.
Graduate School of Engineering, COPPE, Federal University of Rio de Janeiro

Magnani, H. O.
Department of Civil Engineering, Federal University of Santa Catarina

Ehrlich, M.
Graduate School of Engineering, COPPE, Federal University of Rio de Janeiro

KEYWORDS embankment, reinforcement, soft clay, stability

ABSTRACT The paper presents the behaviour of two reinforced test embankments built on a normally consolidated soft clay deposit underlying a top sand layer. The embankments were constructed close to undrained conditions in about 50 days. The embankments were well instrumented, including for measurements of tension forces in the reinforcements. Mobilised tension forces in the reinforcements were shown to increase with embankment height and larger values of tension forces were measured in the embankment built on a shallower clay layer. Critical failure surfaces obtained in limit equilibrium stability analyses were close to the observed field failure surfaces. These analyses used measured reinforcement forces and resulted in Bjerrum correction factors around $\mu = 0.60$ for the two embankments. These analyses considered the measured reinforcement forces and three dimensional effects. Consistent correlations were obtained between embankment loadings, factors of safety, measured reinforcement forces, and inclinometer readings.

1 INTRODUCTION

In order to provide quality data for the geotechnical design and construction control for the future motorway project three test embankments – two reinforced and one unreinforced – were performed (Magnani, 2006). Reinforced mobilized tension forces were monitored during construction using load cells.

The mechanisms involved in the mobilised forces in reinforced embankments on soft clays are discussed in this paper. The analysis is based on two reinforced test embankments taken to failure on a soft normally consolidated clay deposit. Emphasis is given to the results of the measured reinforcement forces and the relationship of these forces with the inclinometer measurements and computed factors of safety. This paper summarizes the behaviour of the two reinforced embankmens discussed in greater detail by Magnani et al. (2009).

2　TEST EMBANKMENT

2.1　Soils, reinforcement and drains

Table 1 presents a summary of the main geotechnical characteristics of Florianópolis soft clay. The geotechnical parameters of this very soft clay are typical of the Brazilian coastal marine deposits. The soft clay under the test embankment presents the behaviour of normally consolidated clay deposit owing to the construction of the hydraulic fill working platform six years before the construction of the test embankments.

TABLE 1　GEOTECHNICAL PARAMETERS OF FLORIANÓPOLIS SOFT CLAY

Parameter	Value
Water content w (%)	100–170
Liquidity index w_L (%)	105–165
Average plasticity index I_P (%)	80
Bulk weight γ (kN/m^3)	13.2–14.2
Voids ratio e	2.9–4.5
Compression ratio $C_c / (1 + e_o)$	0.30–0.45
Coefficient of vertical consolidation c_v	0.7–1.0×10^{-8}
Sensitivity (vane)	3–6

The test embankments were conceived with different features in order to yield relevant data for the motorway construction. Test embankment 1, TE1, was conceived with vertical drains and reinforcement, as is generally adopted in motorways, and test embankment 2, TE2, was conceived with reinforcement only. Test embankment 3, TE3, without drains or reinforcement, was also built, but this is not addressed in the present paper (see Magnani, 2006). Table 2 presents the main features of the two reinforced test embankments. For the purpose of the analyses carried out in the present paper, the influence of vertical drains was relatively small.

Note that although the embankments were located quite close to one another, the thicknesses of the underlying soft clay layers were quite different.

TABLE 2　MAIN FEATURES OF THE THREE TEST EMBANKMENTS

	TE1	TE2
Reinforcement	Polyester Stabilenka -200×45 kN/m; J = 1700 kN/m	
Vertical drains	Colbondrain CX 1000, triangular array, 1.30 m spacing	No drains
Clay thickness (m)	8.2	5.6
Working platform thickness (sand hydraulic fill) (m)	1.7	1.8

Continuous piezocone undrained strength profiles at the centre of each embankment are presented in Fig. 1.

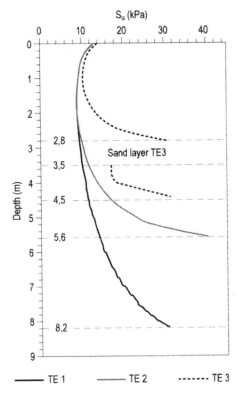

FIG. 1 *Undrained strength profiles at test embankments*

2.2 Geometry and instrumentation

Fig. 2 presents the cross section of embankment TE1 (before and after failure) with slope 1(V):1.5(H). The two test embankments TE1 and TE2 were fully instrumented for vertical and horizontal displacements pore pressures and also tension forces at the reinforcement. For the present paper the relevant

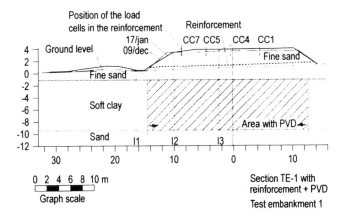

FIG. 2 *Embankment geometry and instrumentation analysed*

measurements are horizontal displacements and reinforcement forces and these are shown in Fig. 2 for test embankment TE1.

Test embankment TE2 had essentially the same geometry, position of load cells, and position of inclinometers as TE1. As mentioned before, TE2 did not have vertical drains and the soft clay layer was less thick. The overall geometry allowed plane strain conditions and induced failure in the central region. The direction of the failure was naturally induced by the gentle inclination of the embankment base. The test embankments were constructed in 50 days.

Reinforcement forces were measured at four points in the reinforcement (see Fig. 2). The load cells used to monitor reinforcement forces (Magnani, 2006) have been used in a number of studies carried out at COPPE in recent years (e.g., Saramago and Ehrlich, 2005).

3   EMBANKMENT BEHAVIOUR

3.1 Horizontal displacements

The inclinometer measurements in embankment TE1 are shown in Fig. 3. These measurements are related to the inclinometers located at the embankment toe which presented the largest measured values, as expected. The results presented in Fig. 3 are horizontal displacements δ_h (Fig. 3) and vertical deviation θ_v (Fig. 4). The vertical deviation is defined as the increment in horizontal displacement $\Delta\delta_h$ divided by the distance between the measured

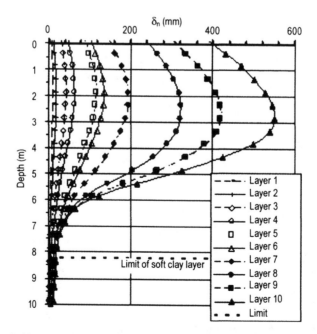

FIG. 3 *Horizontal displacements versus depth for each embankment layer*

points Δz, that is $\theta_v = \Delta\delta_h/\Delta z$. Horizontal displacements were greater in embankment TE1 where a thicker clay layer occurs (Magnani, 2006). The pattern of displacements for embankment TE2 was similar to embankment TE1.

Fig. 5 presents the results of maximum vertical deviation θmax versus time for the inclinometers of embankment TE2. The instant of greatest change in the inclination in the slope curves θmax versus time is related to the construction of the ninth layer (see vertical lines in Fig. 4). This threshold point virtually coincides in all the inclinometers and indicates the onset of failure of the clay foundation. A similar trend was observed in embankment TE1.

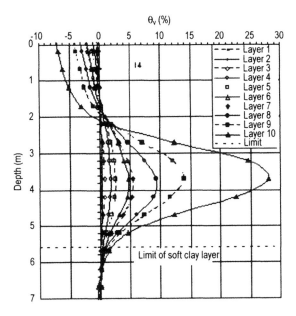

FIG. 4 *Vertical deviation versus depth for each embankment layer*

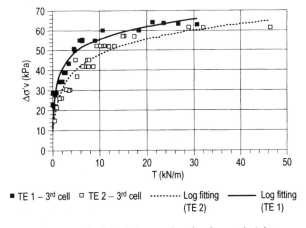

FIG. 5 *Maximum vertical deviation and embankment height versus time*

In both embankments the eighth layer of construction may be considered the last stable condition.

Cracks appeared in both embankments at the ninth layer of construction but failure was more clearly seen in the tenth layer in both embankments.

Embankment TE2 presented a systematically larger vertical deviation than embankment TE1. This is attributed to the smaller thickness of the clay layer underneath the surface of embankment TE2 compared with the corresponding one of embankment TE1, thus generating a larger maximum vertical deviation.

3.2 Measured and estimated reinforcement forces

The values of forces measured at the four load cells installed in the reinforcement in embankments TE2 are shown in Fig. 6 for each fill layer. Note that the maximum reinforcement forces just after construction measured in embankment TE2 verified in the central region of the embankment.

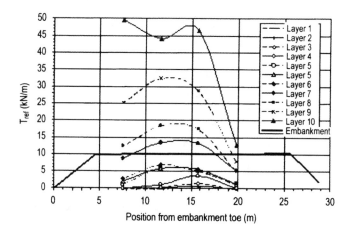

FIG. 6 *Measured reinforcement forces for each embankment layer*

Values of the reinforcement force T measured for different embankment loading conditions are shown in Table 3. Measured T values for the two embankments (TE1 and TE2) are quite close. Geotextile reinforcement strains? a computed from measured T values and reinforcement modulus $J = 1,700\,kN/m$ are less than 0.5% in service conditions and in the range 0.6–1.1% at the onset of failure, reaching 2.4–3.0% when the embankment started to crack.

Fig. 7 presents the values of the maximum reinforcement forces T measured with time and embankment height h for embankment TE1. Similar results were obtained for embankment TE2 (Magnani, 2006).

As expected, the measured values of T increase with h and, just after the ninth fill layer (vertical lines in Fig. 7), which may be considered the moment

TABLE 3 MAIN FEATURES OF THE TWO EXPERIMENTAL REINFORCED EMBANKMENTS

Embankment condition	Measured reinforcement forces T (kN/m)	
	TE1	TE2
Service behaviour ($F_s \approx 1.4$)	4	7
Onset of failure	11–20	18
Embankment cracking	40	50
Maximum value after failure	—	70

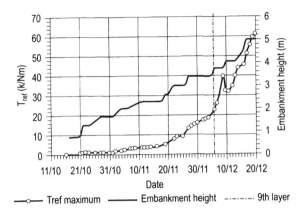

FIG. 7 *Reinforcement forces and embankment height versus time for test embankment TE1*

at which failure occurred in both embankments, the values of T present a greater rate of increase.

Measured tension in the reinforcement T versus the applied embankment load $\Delta\sigma_v$ is presented in Fig. 8 for both embankments. The values of T

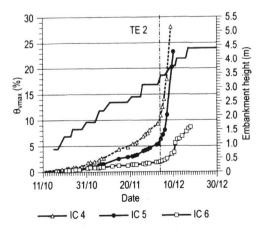

FIG. 8 *Embankment loading versus reinforcement forces measured at the third load cell*

presented are those measured at the third load cell located in the centre of the embankment where the cracks appeared. It may be observed that greater forces T were mobilized in embankment TE2 than in embankment TE1. This may be attributed to the shorter distance between the failure surface and the base of the clay layer for embankment TE2 compared with TE1. Therefore, the gradient of horizontal displacements is greater and so are the vertical deviation values, as explained earlier and thus the T values are also greater.

4 STABILITY ANALYSES OF THE TEST EMBANKMENTS

4.1 Parameters and hypotheses

Stability analyses of the two test embankments were carried out based on the parameters shown in Table 5. The stability analyses carried out for each embankment layer used the values of the forces T measured in the load cells (see Fig. 5).

TABLE 4 PARAMETERS ADOPTED IN STABILITY ANALYSES

	Embankment and working platform	Soft clay
Bulk weight, kN/m^3	17.5	14.0–15.0
Strength parameters	c = 0; ϕ' = 33.8^o	see Fig. 1
Reinforcement forces	Values measured in each load cell in each loading stage	

The analyses took into account the deformed embankment geometry of each layer due to the overall embankment deformation. The estimations of the gain in strength were not considered as this is the usual procedure for stability analyses of one-step constructed embankments.

4.2 Computed factors of safety

Stability analyses were carried out using the modified Bishop method, as the monitoring data and field evidence indicated circular failure surfaces. The field evidence indicated that the embankment failures were three-dimensional (3D), and thus the correction proposed by Azzouz et al. (1983) was implemented. The Bjerrum correction factor μ (Bjerrum, 1972) obtained by back-analysis was close to 0.60.

The variation in the factor of safety (using μ = 0.60) with the embankment loading $\Delta\sigma_v$ is shown in Fig. 9. It is observed that the factors of safety for the two embankments are quite close in all the loading steps, which suggests again that the two embankments had similar behaviour.

It is also noticed in Fig. 9 that the factors of safety for which the failures occurred are not exactly equal to unity (assuming μ = 0.60 and failure in the ninth layer). The range of μ values which resulted in FS3D = 1.0 is shown in Fig. 10 together with other Brazilian case histories (Sandroni, 1993). The

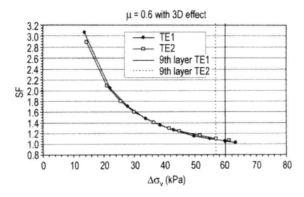

FIG. 9 *Factor of safety versus embankment loading for embankments TE1 and TE2*

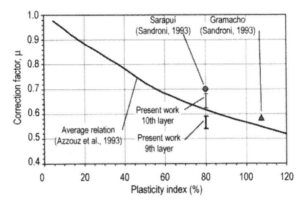

FIG. 10 *Correction factor versus plasticity index: case histories and Azzouz et al. (1983) relationship.*

relevant plasticity index for the present case is Ip = 80%. Values of μ have also been computed for the tenth layer for the two embankments, considering that the failure was identified visually in the tenth layer, although measurements indicated the ninth layer as the failure condition. Fig. 10 also presents the relationship proposed by Azzouz et al. (1983), which is valid for three-dimensional failures, and it is observed that the values obtained in the present study (for the ninth and tenth layers) are consistent with the authors' curve.

5   REINFORCEMENT FORCES, FACTOR OF SAFETY, AND SHEARING BEHAVIOUR

Fig. 11 shows the relationship between the factor of safety and the reinforcement force T. The values of T correspond to the measured values at the load cell located close to the failure surface (third load cell) after the fourth layer of embankment construction (measurements for the first three layers resulted in quite small T values). Note that the values of T increase with the decrease

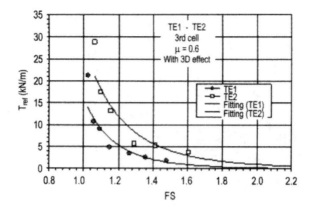

FIG. 11 *Measured reinforcement forces versus factor of safety for test embankments TE1 and TE2*

in the factor of safety and only values of Fs lower than 1.2 have mobilised significant values of T.

Fig. 12 shows the relationship between the inverse of the factor of safety and the maximum angular distortion γ_{max} mobilized *in situ*. Fitting curves are also included in Fig. 11 for the two embankments' data. It is observed that the shapes of these curves are similar to the shape of stress–strain curves measured in undrained triaxial tests in normally consolidated clays. Thus, these data suggest that the shearing behaviours of the two embankments were quite similar. Therefore, the small gain in undrained strength of the clay under embankment TE1 with drains might not be significant enough to markedly differentiate the behaviours of the two embankments. Although the factors of safety and shearing behaviours of the two embankments were similar, higher values

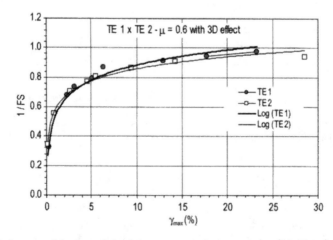

FIG. 12 *Inverse of the factor of safety versus maximum angular distortion for test embankments TE1 and TE2*

of reinforcement forces T were measured in embankment TE2, which appear to be related to the clay layer being thinner than that of embankment TE1.

The relationship between the factor of safety FS, the embankment applied stress $\Delta\sigma_v$ and mobilized reinforcement tension T shown in previous figures may be summarized conceptually as shown in Fig. 13. The maximum reinforcement tension Tlim shown in Fig. 13 corresponds to a foundation fully plastified.

The benefitial influence of the reinforcement stiffness J (not studied in the present paper) is seen in Fig. 13. For the same embankment load $\Delta\sigma_v$ a stiffer reinforcement will mobilize a higher T value, thus a higher factor of safety is reached. Alternatively, for the same T value the stiffer reinforcement results in a higher FS.

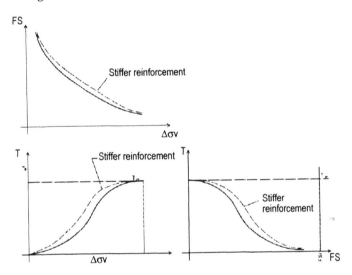

FIG. 13 *Conceptual behavior of reinforced embankments on soft clays*

6   CONCLUSIONS

Two reinforced test embankments, TE1 and TE2, were built on soft normally consolidated clay layers. Embankment TE1 was provided with vertical drains to simulate the conditions of a real engineering project.

Mobilised tension forces T in reinforcements were shown to increase with the embankment height. Larger values of T were measured in embankment TE2, built on a shallower clay layer, which also presented larger values of vertical deviation measured by the inclinometers.

The curves of vertical deviation versus time showed changes in the curvature that coincided with the instant in which cracks appeared in both embankments. A change of curvature, although with a smaller gradient, was also noticed in the curves of maximum reinforcement forces versus time.

The curves of embankment loading versus maximum vertical deviation were quite similar for the two embankments, and were also similar in shape to stress–strain curves of undrained triaxial tests performed in normally consolidated clays. The two reinforced embankments presented similar factors of safety in all loading stages, although one of them was built over a thicker layer and was provided with vertical drains. The reinforcement loads increased with the decrease in the factor of safety. A conceptual behaviour of reinforced embankments was proposed based on the results obtained in the present study.

REFERENCES

Azzouz, A. S., Baligh, M. M. and Ladd, C. C. (1983). Corrected field vane strength for embankment design. *Journal of Geotechnical Engineering ASCE*, v. 109(5), p. 730.

Bjerrum, L. (1972). Embankments on soft ground. *Proceedings of the Special Conf. on Performance of Earth and Earth-Supported Structures*, ASCE, vol. 2, pp. 1–54, Purdue, New York.

Magnani, H. (2006). *Behaviour of reinforced embankments on soft clays taken to failure* (in Portuguese), DSc thesis, COPPE/UFRJ, Rio de Janeiro, RJ, Brazil.

Magnani, H. O., Almeida, M. S. S. and Ehrlich, M., (2009). Behaviour of two reinforced test embankments on soft clay. *Geosynthetics International*, 2009, 16(3), 127–138.

Sandroni, S. S. (1993). On the use of vanes tests in the design of embankments on soft clays (in Portuguese). *Solos e Rochas*, 16(3), 207–213.

Saramago, R. P. and Ehrlich, M. (2005). Physical 1:1 scale model studies on geogrid reinforced soil walls. *Proceedings of the Sixteenth International Conference on Soil Mechanics and Geotechnical Engineering*, vol. 2, pp. 1405–1408, Osaka, Japan.

The Role of *In Situ* Testing on Interactive Design on Soft Clay Deposits

Schnaid, F., Consoli, N. C., Dalla Rosa, F. and Rabassa, C.
Federal University of Rio Grande do Sul – Brazil

KEYWORDS soft clay, *in situ* tests, soil parameters, interactive design

ABSTRACT The ability to predict the performance of geotechnical structures – in particular on soft clay deposits – consists in one of the major challenges confronting engineers. A key factor affecting predictions is related to our capacity to estimate realistic soil parameters which has improved recently with the advent of modern testing techniques and more rigorous methods of analysis. These developments have been incorporated into engineering practice and among the several different techniques a combination of piezocone tests and laboratory oedometer test from undisturbed samples is now considered as minimum requirement in any routine site characterization in soft clay deposits. This paper addresses the current experience by given a brief review of site characterization procedures and *in situ* testing interpretation and their role on interactive design. Examples reporting the Brazilian practice are discussed.

1 INTRODUCTION

Site characterization and *in situ* test interpretation have been evolving from basic empirical recommendations to a sophisticated area demanding a through knowledge of material behaviour and numerical modeling (e.g. Schnaid, 2005). These developments have posed new challenges to the geotechnical community by increasing the need to narrow the gap between specialized researchers and practicing engineers, who may not be familiar with new detailed methodologies but have the need to incorporate new concepts into design to provide safer and more economical projects.

A variety of *in situ* tests is commercially available to meet the needs of geotechnical engineers. These existing field techniques can be broadly divided into two main groups (after Schnaid, 2005):

a. *Non-destructive* tests that are carried out with minimal overall disturbance of soil structure and little modification of the initial mean effective stress during the installation process. The non-destructive group comprises seismic techniques, pressuremeter probes and plate loading tests, a set of tools that is generally suitable for rigorous interpretation of test data under a number of simplified assumptions.

b. *Invasive, destructive tests* were inherent disturbance is imparted by the installation of the probe into the ground. Invasive-destructive techniques comprise SPT, piezocone and dilatometer. These penetration tools are robust, easy to use and relatively inexpensive, however, the mechanism associated with the installation process is often complex and therefore a rigorous interpretation is only possible in few limited cases.

Assessment to the properties of clay requires a combination of techniques that should be carefully planned to provide shear strength, stress history and compressibility measurements from a detailed site investigation program. In a routine bases, the minimum standard of work requires piezocone testing data and laboratory oedometer tests to be combined. Specific recommendations are now available to derive undrained shear strength, soil stiffness, stress history and consolidation coefficients from piezocone data (Jamiolkowski et al., 1985; Lunne et al., 1997; Meigh, 1987; Robertson and Campanella, 1988; 1989; Yu, 2004; Schnaid, 2005; Mayne et al., 2009). Oedometer tests give compression measurements and tune the consolidation coefficient and stress history assessed by the piezocone.

Despite the experience gathered in past decades, two frequent questions are often raised by practicing engineers: how to link theory to practice in routine geo-structure design and how reliable is the standard of practice in onshore, nearshore and offshore operations. These questions are addressed in this paper and are used to illustrate the current practice in Brazil and the experience on interactive design.

2   SITE CHARACTERIZATION

The so-called interactive design (or observational method) is recommended when prediction of geotechnical behavior is difficult. Under difficult conditions Eurocode 7 states: "it can be appropriate to apply the approach known as observational method, in which the design is reviewed during construction". Since earthfill construction on soft ground generally involves high risk and it is time dependent – soil properties and therefore factors of safety change over a period of time – interactive design is always an appealing approach. It allows the performance of a geo-structure to be continuously checked and the design model and its constitutive parameters to be continuously updated and re-calibrated as required, whilst future performance is re-assessed (e.g. Negro Jr. et al., 2009).

Among the main uncertainties concerned to engineering design, assessment to boundary conditions and soil properties are as fundamental. The real geometry of a problem needs considerations to soil stratification and its 1D, 2D or 3D representations. The actual ground conditions can be achieved by a detailed and planned site investigation. A wide range of dedicated equipments with highly specialised workmanship is now available worldwide comprising very light weight rigs, trucks and vessels that support remote controlled operations. On-shore operations are often performed on vehicles that provide quick, cost-effective mobile platforms capable of high production rates.

Several companies are able to provide robust, off-road 100 to 200 kN trucks that may have difficulties in accessing soft ground locations. Operations in soft ground can be performed using lightweight rigs that are mounted on a trailer or mini crawlers with compact dimensions, supporting single-cylinder or two hydraulic cylinders units. These light units are often built locally to meet economical and technical constraints.

Marine geotechnical off-shore investigations can be performed from dedicated drillships equipped for geological survey, geotechnical stratigraphy and reservoir characterization. These ships are often equipped with seabed units that can perform tests mechanically with the unit submerged and placed onto the seabed where the test is performed automatically according to prescribed procedures. Contact from the surface with the seabed unit is by means of acoustic data modem or by a bypass cable.

Acceptable working conditions for near-shore operations are more difficult to set due to limitations of available equipment at reasonable cost. Knowledge of seabed soils is essential in near-shore operations, since many coastal regions are developed in soft compressible clays with low bearing capacity where there is continuous infrastructure development. If the water depth is less than about 20 m, geotechnical operations may be carried out as onshore by using ordinary drilling from jack-up platforms or alternatively using compressed air bells. Anchored barges can be used in sheltered areas, although barge motion may affect the geotechnical quality. In greater water depths, or if wave, current or wind conditions are severe, it is necessary to employ vessels fitted with anchoring or dynamic positioning systems used to deploy submerged seabed units.

Fig. 1 illustrates the characteristics of a submerged seabed unit especially designed for a 20 m depth investigation in southern Brazil. The system carries a hydraulic rig that is deployed into the sea using a single steel wire and the contact from the surface with the seabed unit is by means of a bypass cable. CPTU tests can be performed and undisturbed samples can be retrieved. A series of CPTU tests carried out along the axis of the breakwater (typical result in Fig. 2) allowed the characterization of a complete soil profile (Fig. 3) which has been used to define the stratification of the site at the design stage.

3   SOIL PROPERTIES: COEFFICIENT OF CONSOLIDATION

Under interactive design, rate and degree of consolidation of soft clay deposits can be constantly assessed to reduce uncertainties rising from soil parameters, stress history and boundary conditions (stratification and geometry). Take a saturated soil mass subjected to a given load, where there is an increase in pore water pressure followed by consolidation. This process takes time before the soil mass can reach a steady state. Under primary consolidation the rate of settlement is governed by Darcy's law and hydraulic conductivity and is

FIG. 1 *Submerged seabed unit (Courtesy CBPO, PEDRASUL, CARIOCA & IVAÍ)*

conveniently expressed by the coefficient of consolidation. The coefficient of consolidation C_v is not a soil parameter but a function of the deformation modulus M and the vertical hydraulic conductivity k_v:

$$C_v = \frac{kM}{\gamma_w} = \frac{k\sigma_v'(1 + e_o)}{0.434\gamma_w C_c}$$

where M is the deformation modulus defined as $d\sigma_v'/d\varepsilon_v$ in one-dimensional compression. Note that the coefficient of consolidation is larger in the over-consolidated domain than in the normally consolidated domain, where the hydraulic conductivity decreases with void ratio and the deformation modulus is lower. Even in this overconsolidated domain, as the bulk modulus is generally not constant, C_v is not constant either (e.g. Leroueil and Hight, 2003).

Several methods have been proposed for determining coefficient of consolidation from laboratory oedometer consolidation curves obtained from one-stage loading: Casagrande (log t) method; Taylor ($\sqrt{t}$) method; rectangular hyperbolic fitting method (Sridharan and Sreepada Rao, 1981), among others. Continuous loading oedometer tests, CRS, can also be used to determine C_v as a function of strain. Alternatively the coefficient of consolidation can be assessed from *in situ* tests, preferably from piezocone dissipation tests (Levadoux and Baligh, 1986; Teh and Houlsby, 1991; Burns and Mayne, 1998). A brief summary of these methods is given by Schnaid (2009).

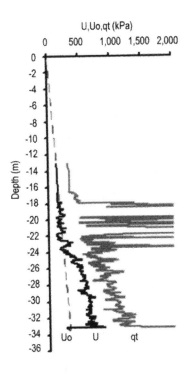

FIG. 2 *Typical piezocone profile on seabed soft clay overlain silty-sand layer*

BREAKWATER EAST

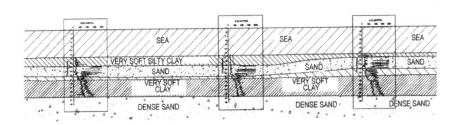

FIG. 3 *Profile of nearshore breakwater construction*

A question that emerges in every design project is how representative is the coefficient of consolidation determined from both laboratory and *in situ* tests. Two examples case studies are presented below to highlight the reliability in assessing the characteristics of consolidation of clay deposits for both onshore and offshore applications.

Schnaid et al. (2001) adopted the interactive design approach to study the geotechnical performance of an embankment built over a soft clay deposit at the new International Airport in the city of Porto Alegre, southern Brazil. In order to accelerate the consolidation process, prefabricated vertical drains (PDV) were installed in the area together with a temporary surcharge. Construction was preceded by a comprehensive site investigation comprising both laboratory and *in situ* testing, and was followed by the installation of field instrumentation with settlement plates and piezometers. The analysis and monitoring of the embankment indicate that design methods reproduced essential features of the consolidation process and that the performance of the drains in the compressible layer was considered adequate.

Values for the coefficient of consolidation for the normally consolidated range are summarized in Figs. 4 and 5. Observational methods such as Asaoka (1978) and Magnan and Deroy (1980) have been used to deduce the consolidation values from settlement observations. Complementary, finite element curve fitting of settlement × time data has been adopted for back-analyzing the measured data and to adjust the model parameters.

For the region treated with vertical drains, the *in situ* coefficients of horizontal consolidation in the normally consolidated range are typically twice the values deduced from oedometer test with radial drainage near the surface. At greater depth the values coincide. The C_h values deduced from piezocone dissipation tests, corrected to represent the normally consolidated domain (Jamiolkowski et al., 1985), are typically 50% lower than values obtained by back-analysis. It is important to recall that *in situ* and laboratory data are representative of the normally consolidated range whereas values from the

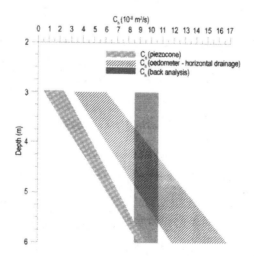

FIG. 4 *Horizontal coefficient of consolidation C_h (Porto Alegre International Airport)*

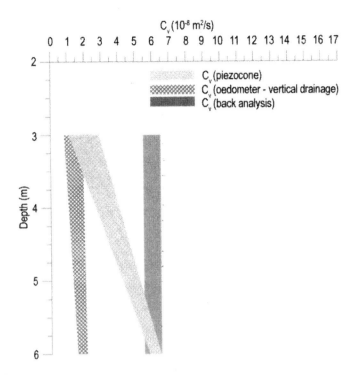

FIG. 5 *Vertical coefficient of consolidation C_v (Porto Alegre International Airport)*

back-analysis represent an average response that may be partially affected by some overconsolidation near the surface.

For the condition of one dimensional consolidation (no vertical drains used) both the oedometer cell and the piezocone values underestimate the values of C_v from back-analysis (Fig. 5). The difference reduces with depth which suggests that discrepancies are partially due to stress history (the soil is slightly overconsolidated at lower depth).

Another example of interactive design describes the stage construction of a 20 m high marine breakwater laying over a 20 m thick soft clay deposit at the Rio Grande Harbour in southern Brazil. The major risk associated to the project is during the elevation of the breakwater from the level +2 m above sea bed to its final +5.5 m level (minimum factor of safety), which prompted a careful prediction of the breakwater closure. At this stage (+2 m level) a finite element curve fitting of settlement × time data has been adopted for back-analyzing all measured data, comprising settlements, horizontal displacements and pore-water pressure. Partial drainage has been considered in every previous construction stage until measured and predicted data have been adjusted. This procedure allowed the field coefficient of vertical consolidation for the normally consolidated range to be estimated and to be compared to those measured by laboratory and *in situ* test data (Fig. 6). Values are shown to reduce with depth reflecting the soil stratum at the site: a superficial very soft silty layer, recently deposited, and a sandy-clay layer up to 25 m overlain a thick soft clay layer (as shown in Fig. 3). Whereas the values of C_v from back-analysis are generally greater than laboratory and *in situ* test values for the superficial layers, good agreement is observed for all tests at the soft clay layer.

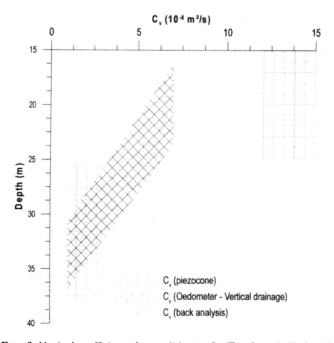

FIG. 6 *Vertical coefficient of consolidation C_v (Rio Grande Harbour).*

As a final remark, it is recognized that the coefficient of consolidation remains as one of the most difficult geotechnical parameters to determine, but our ability in producing realistic estimation from laboratory and *in situ* test data has improved considerably.

4   CONCLUSIONS

Operational risk associated to geo-structures on soft clay deposits are strongly related to uncertainties about the properties of soils at the site. Geological, hydro-geological and geotechnical information should be acquired by sufficient investigations to reduce these risks to a minimum so that the structure can be safely designed and built. Since earthfill construction on soft ground is time dependent, interactive design appears to be an appealing design approach, one that enables the set of constitutive soil parameters to be adjusted and calibrated during construction in order to check all assumptions adopted at early design stages.

Examples given in this paper highlight the usefulness of *in situ* testing data to evaluate the boundary conditions of a given project and to assess the model parameters. In both case studies, soil parameters assessed from laboratory and *in situ* test data where within the range of values derived from the back-analysis of field instrumentation. In these cases interactive design has proved practical in adjusting the model parameters to refine basic design principles. Cases of poor consistence of model data and field observation may require a critical appraisal of the investigation needs and further site investigation even at late construction stages.

ACKNOWLEDGEMENTS

The authors wish to express their gratitude to the Brazilian Council for Scientific and Technological Research/Brazilian Ministry of Science and Technology (CNPq) for the financial support to the research group. Thanks are also due to Odebrecht Group and Geoforma Consultants for supplying part of the field data used.

REFERENCES

Asaoka, A. (1978). Observational procedure of settlement prediction. *Soils and Foundations*, 18 (4), 87-101.

Burns, S. E. and Mayne, P. W. (1998). Monotonic and dilatory pore-pressure decay during piezocone tests in clay. *Can. Geotech. J.*, 35(6): 1063-1073.

Jamiolkowski, M., Ladd, C. C., Germaine, J. T. and Lancellotta, R. (1985). New developments in field and laboratory testing of soils. *Proc. 11th International Conference on Soil Mechanics and Foundation Engineering*, San Francisco, 1, 57-153.

Leroueil, S. and Hight, D. W. (2003). Behaviour and properties of natural and soft rocks. *Charact. and Engng. Properties of Natural Soils*, Balkema, 1: 29-254.

Levadoux, J.-N. and Baligh, M. M. (1986). Consolidation after undrained piezo-cone penetration. *J. of Geotech. Engng., ASCE*, 112(7): 707-726.

Lunne, T., Robertson, P. K. and Powell, J. J. M. (1997). Cone penetration testing in geotechnical practice, *Blackie Academic & Professional*, 312p.

Magnan, J. P and Deroy, J. M. (1980). Analyse graphique des tassements observes sous les ouvrages. Bull. *Liaison Laboratoire des Ponts e Chaussés*, Paris, 109, 9-21.

Mayne, P. W., Coop. M. R., Springman, S. M., Huang, A. B. and Zornberg. J.G. (2009). Geomaterial behavoir and testing. *Proc. 17th International Conference on Soil Mechanics and Geotechnical Engineering*, Alexandria, 4, 2777-2872.

Meigh, A. C. (1987). Cone penetration testing – methods and interpretation, *Construction Industry Research and Information Association*, CIRIA, London., 141 p.

Negro Jr., A., Karlsrud, K. Srithar, S., Ervin, M. and Vorster, E. (2009). Prediction, monitoring and evaluation of performance of geotechnical structures. *Proc. 17th International Conference on Soil Mechanics and Geotechnical Engineering*, Alexandria, 4, 2930-3005.

Robertson, P. K. and Campanella, R. G. (1988). *Guidelines for geotechnical design using CPT an CPTU*, University of British Columbia, Vancouver, Department of Civil Engng., Soil Mech. Series 120.

Robertson, P. K. and Campanella, R. G. (1989). *Design Manual for use of CPT and CPTU.* University of British Columbia, Vancouver, BC.

Schnaid, F., Nacci, D and Mililitsky, J. (2001). *Salgado Filho International Airport: Geotechnical Infrastructure*, Sagra Luzzatto, 222p. (in Portuguese)

Schnaid, F. (2005). *Geocharacterisation and properties of natural soils by* in situ *tests: a summary.* Ground Engineering, 38 (9), 23-24.

Schnaid, F. (2009). In Situ *Testing in Geomechanics.* Taylor & Francis, 329p.

Sridharan, A. and Sreepada Rao, A. (1981). Rectangular hyperbola fitting method for one-dimensional consolidation. *Geotechnical Testing Journal*, 4 (4), 161-168.

Teh, C. I. and Houlsby, G. T. (1991). An analytical study of the cone penetration test in clay. *Géotechnique*, 41(1): 17-34.

Yu, H. S. (2004). The James K. Mitchell Lecture: *In situ* testing: from mechanics to prediction. 2nd *Int. Conf. on Site Charact.*, Milpress, Porto, 1: 3-38.

Vacuum consolidation, vertical drains for the environment friendly consolidation of very soft polluted mud at the Airbus A-380 factory site

Varaksin, S.

Menard, Nozay, France, serge.varaksin@menard-mail.com

KEYWORDS heavy metals, The Menard Vacuum®, concept, Vacuum process, sprinkling method, G.C.C. (geotextile confined columns), impervious wall, very soft clay

ABSTRACT The extension of the existing Airbus factory in Hamburg, Germany on a former sand quarry filled with recent alluvium required techniques avoiding the displacement of the mud into the Elbe River and the control of the pore water, containing amonium and heavy metals. The Menard Vacuum®consolidation technique provides stability and accelerated consolidation of the very soft muds without any failure risk. Zero emissions vacuum pumps were utilized and the controlled consolidation liquid was channeled towards a sewage treatment unit. Six years of monitoring of the taxiways confirmed the success of the method.

1 FOREWORD

The four years term of the TC 17 has concentrated on two themes: concept and soil parameters, and also rigid inclusions.

This paper bridges the "concept and parameters" theme with the next term (2009-2013) of the TC 17 being sustainable technologies. The contribution of Dr. Jian Chu in his working group (B2) "design and construction using preloading and vertical drains" and "vacuum consolidation" (B3) by J. Kirstein should be read in conjunction with this case history presentation. They can be downloaded from the TC 17 site (http://www.bbri.be/go/tc17).

This case history had also very particular conditions: not only was the concept, the proposal of the contracting teams, made by a specialized company of Marine Works (J. Moebius) and soil improvement (Menard) but also at award stage, criteria of long term behaviour were not defined by the owner and a contractual approach had to be defined allowing variations in schedule and long time residual settlement guarantees.

The only available information were a comprehensive detailed soil investigation, the specifications of the hydraulic fills, the final elevation and of course, the stringent requirements to avoid any mud creep into the adjacent Elbe River also together with a near zero emission (gas and consolidation liquid outflow) requirement.

2   PROJECT DESCRIPTION AND ENVIRONMENTAL CONSTRAINTS

Airbus industries has selected a site, adjacent to the existing factory and the existing runway, for the unloading, assembling, painting and final delivery of the future mega plane, the A-380 (Fig. 1).

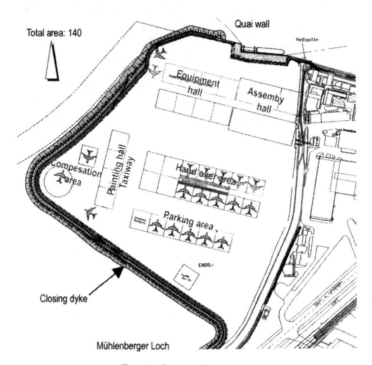

FIG. 1 *General Site layout*

This location covering approximately 140 hectares is a former sand quarry exploited in the first half of the 20th century and since then, abandoned and presently silted in by Elbe River alluvium and thus heavily polluted.

Environmental consideration required construction methods excluding all lateral movements of mud deposits, and that all water expulsion from the consolidation of the muds should be treated due to the high concentration of Ammonium and heavy metals.

3   SUBSOIL CONDITIONS

The site, contiguous to the existing Airbus plant is under tidal influence.

Average variations of the tide is from (−2 m) to (+2 m). In certain periods of conjunction of high tides and East winds, the highest water elevation can reach +6.50 m.

TABLE 1 SUBSOIL CHARACTERISTICS

Soil type	Water content	Density	Shear strength		Deformation Modulus	Coefficient of consolidation	Coefficient of secondary consolidation
				(under $\sigma_z = 100\,\text{kN/m}^2$)			
	W (%)	γ/γ' (kN/m^3)	$\varphi'(°)/c'$ (kN/m^2)	C_u (kN/m^2)	E_S (MN/m^2)	C_V (m^2/year)	$C\alpha$ (-)
Mud	142	13/3	20/0	0.5-5	0.8	0.35	0.03
Young clay	119	14/4	20/0	2-10	0.9	0.35	0.03
Clay	70	15/5	17.5/10	5-20	1.5	0.5	0.02
Peaty clay	139	14/4	20/5	5-20	0.9	0.4	0.03
Peat	240	11/1	20/0	5-15	0.5	≥ 0.4	0.04

The table (1) summarizes subsoil characteristics. The thickness of compressible layers varies from 5 to 14 m and a surface mud layer of 3 to 12 m thickness excludes all access to the site.

Accessibility by floating flat bottom barges is also limited to $1\frac{1}{2}$ hour per tide. The cohesion of surface mud layers measured by special wide blades vane tests resulted in the range of $C_u \approx 0.5\,\text{kN/m}^2$ equipment neither movement nor fill in excess of 30 cm could be envisaged without failure.

Considering a final elevation of +5.5 m and excluding lateral movement, theoretical vertical deformation under the fill weight did range from 2.5 to 4.0 m; secondary deformation had to be added if organic deposits were present.

4   GENERAL CONCEPT OF THE ENGINEER.

The basic design of the Engineer resulted in 3 tenders, launched by the developer of the site, who is a subsidiary of the Port of Hamburg.

Tender n° 1

This tender called for a permanent quay wall construction for the purpose of unloading plane elements transported by barge and a peripherical temporary sheetpile wall, with the purpose to contain the muds and isolate the site from tidal influence.

Tender n° 2

Those works consisted in raising the water level inside the sheetpile wall to elevation +4 m, sprinkle 3,000,000 m^3 of sand in thin layers.

The sand was to be reclaimed from an island in the Elbe River, 5 km away, and a site in the North sea at 180 km distance of the works.

The sand sprinkling operations were to be performed in 30 cm height single layers, up to the elevation of +3.0 m. This procedure would prevent mudwaving due to bearing capacity failure. The next phase of works would consist in a groundwater lowering to the levation of +0.7 m allowing a suitable working platform for very light equipment. Vertical drains were to be installed in non structural areas and Vacuum consolidation (Menard Type) in the structural areas (204,000 m²), where pavement and buildings were to be constructed.

The aim of the Vacuum process was to allow filling operations and reach the deformation criteria in a very short time excluding any risk of failure.

The fill total quantity of sand was thus designed to minimize additional quantities of sand to compensate for consolidation settlement.

Tender n° 3

Those works would consist in the construction of a permanent dyke on the consolidated grounds within the closing dyke and removal of the temporary sheetpiles.

5   ALTERNATE CONCEPT BY THE SPECIALIST CONTRACTOR

In 1994, the harbor of Lübeck, located approximately 80 km Northeast of Hamburg, accepted new methods of backfilling of an old harbor by using cohesive polluted dredge spoil, covering it with sand using a sprinkling method developed by the "Moebius" company (a local marine works contractor) and consolidating it over an extremely short period by the Vacuum process as developed by Menard.

The engineer of the Airbus site retained those techniques in their concept as described in Fig. 2a.

The Moebius–Menard group of companies presented an alternative to those tenders that would avoid the construction of the temporary sheetpile wall. This alternative required the construction of the permanent dyke of tender n° 3 within the tender n° 1 period (8 months) and thus ahead of the yearly high tide period that would eventually destroy the works if the dyke was not closed (Fig. 2b).

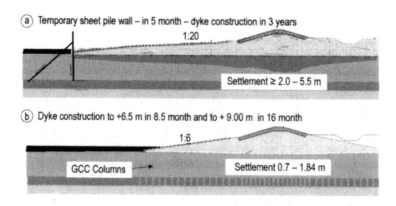

FIG. 2 *(a) Basic design; (b) Alternate concept of Moebius–Menard*

The technique of G.C.C. (geotextile confined columns) developed by the Moebius company was proposed, being performed downwards from floating equipment. This technique should insure the stability of the closing dyke and avoid lateral mud displacement in the adjacent Elbe River, as well as reduce and accelerate deformation of the subsoil with time (Fig. 3).

This technique was demonstrated previously on a test site within the Port of Hamburg and allowed construction and raising of the closing dyke in 3 successive loading increments over a period of 3 months.

FIG. 3 *Dyke construction for the polder formation*

The vertical drains and Vacuum consolidation would allow the subsoil inside the dyke to consolidate within the allocated schedule and criteria.

As mentioned previously, the leasing of the land by Airbus Industries to the owner, the Port of Hamburg not being finalized, the alternate contractors concept was accepted and contracted based on quantities to be defined by a contractual calculation method (TARAO) as works progress, taking into account actual sand delivery, the residual settlement criteria and their time of validity, only defined at a later stage.

6   METHOD STATEMENT

After the formation of the closing dyke, the water is raised to an elevation of approximately +3 m (Fig. 4) to allow flotation of dredging equipment.

A special sprinkling pontoon is manufactured (Fig. 5) enabling the filling of extremely thin sand layers. Indeed, cohesion of the muds being as low as $0.5 \, kN/m^2$, only a few centimeters layers are filled at once each time, avoiding mud waving.

Horizontal drains are placed to provide for dewatering and a working platform for the performing of vertical drains and vacuum consolidation.

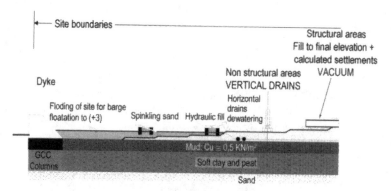

FIG. 4 *Method statement*

FIG. 5 *Sprinkling pontoon*

7   SOME GENERAL ASPECTS OF THE G.C.C. (GEOTEXTILE CONFINED COLUMNS)

The technique consists in driving or vibrating a 80 cm diameter steel casing to the bearing soil, placing a seamless cylindrical closed bottom, Geotextile "sock", with strength ranging from 200 to 400 kN/m, filling it by sand (Fig. 6).

Design and dimensioning have been published by Raithel et al. (2002), where analytical and numerical methods have been proposed.

8   SOME GENERAL ASPECT OF THE VACUUM PROCESS (MENARD TYPE)

The Vacuum consolidation is an alternative to preloading by surcharge associated to vertical drains. The general principle is presented in Fig. 7.

FIG. 6 *G.C.C. Installation process*

The particularity of the Vacuum process as developed by Cognon (1961) and Cognon et al. (1961) is the dewatering below the membrane, permanently keeping a gas phase between the membrane and the lowered ground water. The P', Q' diagram (Fig. 8) allows to understand the major difference between surcharge and Vacuum (Fig. 7).

The Vacuum process evolves from point A to point E, or below the K_0 line. The surcharge process evolves from point A to point B, or towards the K_F line.

This demonstrates that Vacuum consolidation prevents all lateral movement and even creates the opposite phenomenon of inward movement to the

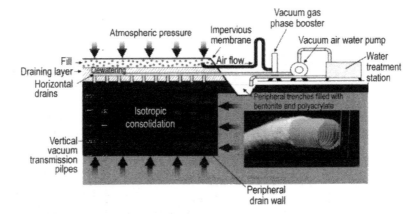

FIG. 7 *Menard Vacuum® process*

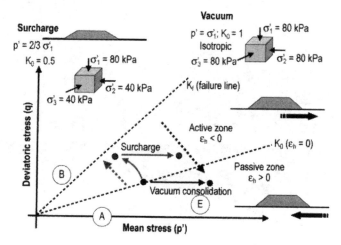

FIG. 8 *Stress path for Vacuum Process under infinite fill load*

treated area. Furthermore, consolidation water is collected at the pumps and directly sent to a sewage station as per specifications.

9 PARTICULAR APPLICATION OF VACUUM AS CONFINEMENT CORSET IN THE ASSEMBLY HALL AREA

The engineer's request to accelerate the program of works and hand over the assembly hall over a period of 8 months, thus before completion of the closing dyke, required special measures over a $130,000\ \text{m}^2$ area.

The "Emergency scheme" shown in Fig. 9 was proposed and performed by the contracting team, included:

Construction of a "mini" dyke on four rows of G.C.C. covered by sand bags to resist a water pressure of 2.5 m height.

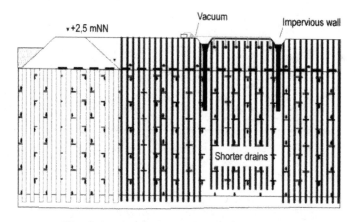

FIG. 9 *Installation of a Menard Vacuum® corset*

Filling of the basin by water to elevation +2.5 m and sprinkling of sand to elevation +2.5 m is performed.

A Vacuum consolidation corset was designed and built to form a retention wall around the future hydraulic fill. This fill had to be placed over a period of a few weeks.

For this purpose, the combined effect of the pre-consolidation of the muds in the Vacuum process and their shear strength increment, including the apparent cohesion in the sand layer under the Vacuum, were taken into account.

The analysis of deformations (including horizontal) and stability by finite elements has demonstrated that sand filling from elevation +2.5 to +9.5 m, with slopes of ¼ would be stabilised by the Vacuum corset and also consolidated according to the client's specifications.

Fig. 10 illustrates the behaviour of the different subsoil layers as function of time, as compared to the theoretical analysis based on laboratory obtained parameters and the contractually defined "TARAO" calculation method.

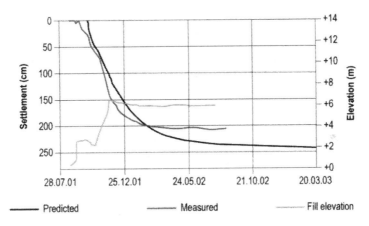

FIG. 10 *North area: measured × predicted settlements*

10 APPLICATION OF THE VACUUM CONSOLIDATION IN THE STRUCTURAL AREAS (TAXIWAY, PARKING AREAS)

The criteria deformation versus time of the taxiway and apron areas being more severe (less than 10 cm post construction settlements) the Vacuum consolidation process was selected to solve the stability problem associated to the filling operations and obtention of the required long-time settlement criteria.

Furthermore, those structures were located over compressible soils extending not only to greater depth, but also containing organic horizons.

A substantial overconsolidation and "aging" had to be obtained to reduce secondary consolidation settlements (calculated to be in excess of

FIG. 11 *Vacuum application on taxiway area*

30 cm over a period of 20 years) within the 10 cm residual settlements under fill, structures and live loads, as required by the specifications.

The installation of the Vacuum system has required the utilisation of high capacity trenching machines, capable to install a geomembrane and a bentonite wall to 8 m depth (Fig. 12).

Indeed, the upper layers consisting of previous "sprinkled" sand had to be totally isolated by a cut off membrane. The deep seated sand seems were

FIG. 12 *Vacuum Taxiway – Construction of the impervious wall*

trenched, mixed with clay of the mud layers and a bentonite injection rail equipped the trenching arm to leave a 40 cm thick impervious cut off.

Fourteen pumping stations, each with one Air-Water, and one Air-Air vacuum pump were installed in a 8 m high, 2.5 m diameter casing, placed on a concrete pad, equipped with a sump pit. This allowed to have the air-water pumps close to the water table and protected from the fill operation during their hydraulic pumping process (Fig. 13).

FIG. 13 *Protection casing layout for Vacuum pumps*

The calculation spreadsheets contractually defined are illustrated in Table 2. Their principle is based on the "target Void ratio" to be reached in each layer to meet the specification and insurance to reach the consolidation under full effective stress in each horizon.

Table 2 also takes into account the water table variation due to tidal action, and the stress change induced as function of the fill settlement below the ground water as settlement occurs.

The site was divided in over 120 zones with a reference boring each. The spread sheet was detailed for each boring according to the:

- load parameters finalized with the client, based on the actual final elevation sand fill delivery schedule and various completion dates;

- initial soil parameters obtained from each boring log and related laboratory tests;

- drainage parameters as drain spacing, and coefficient of horizontal consolidation.

This summarizes the principle of the spread sheet that was submitted for client's approval for every zone to endorse the load parameters and the actual schedule.

TABLE 2 SOIL PARAMETERS – TARAO TYPICAL SPREAD SHEET

Hambourg A380 Ds 59 – XI1017-B62 – (4.19 u.4.20) – + 5.5mNN – Raster: 0.6 m MVC – Bolck Rollwege in FS
Initial elevation:3.5   Fill parameters: Bulk weight (kN/m³): 18   Cohesion (kPa): 0   Friction angle (°): 27.5

Load Parameters Parameters	unit		Initial fill thickness above GWT: 0.50				Initial fill thickness below GWT: 3.1			
			Step 1	Step 2	Step 3	Step 4	Step 5	Step 6	Step 7	Step 8
Fill density	kN/m³	γ_t	18	18	18	18	18	18	18	18
Fill height abovw GWT (incl. exist. fill, beginning of step, no settlement considered)	m		0.5	1	1	2	5.5	5.5	5.23	5.23
Maintain elevation (yes = 1, no = 0) ?			0	0	0	0	0	0	0	0
Fill height above GWT (incl. settl. end of step)	m	H_f	−0.28	−0.38	−1.03	−0.46	2.3	2.26	2	1.86
Fill thickness below GWT (incl. settl., end of step)			3.88	4.48	5.13	5.56	6.3	6.34	6.33	6.47
Fill settlement (end step)	m	W_t	0.78	1.38	2.03	2.46	3.2	3.24	3.23	3.37
Fill width and length	m × m	(Lxl)	Infinite	Infinite	Infinite	Infinite	Infinite	Infinite	Infinite	Infinite
Vacuum pressure	kPa	V	0	0	60	80	80	25.6	25.6	35.2
Coefficient of Vacuum		α_v	1	1	1	1	1	1	1	1
Duration of loading	days	t	25	74	62	57	221	149	39	1,825
Time at beginning of step	days		0	25	99	161	218	439	588	627
Data at beginning of step			22/8/2002	16/9/2002	29/11/2002	30/1/2003	28/3/2003	4/11/2003	1/4/2004	10/5/2004
Fill elevation from GWT (beginning step)	m		0.5	1	1	2	5.5	5.5	5.23	5.23
Fill elevation from GWT (end step)	m		−0.28	−0.38	−1.03	−0.46	2.3	2.26	2	1.86
Elevation at the of step	m		3.22	3.12	2.47	3.04	5.8	5.76	5.5	5.36

TABLE 2 SOIL PARAMETERS – TARAO TYPICAL SPREAD SHEET (CONT.)

Initial Soil Parameters

Layer	unit		1	2	3	4	5	6	7	8
Thickness	m	H	5.1	2.5	1	0.3	1	0.8		
Void ratio Tarao		e_t	3.7	2.7	3.6	3.6	1.4	2.7		
Void ratio		e_0	3.3	2.7	2.7	2.7	1.4	2.7		
Primary consolidation Tarao		C_{ct}	0.95	0.83	0.9	0.9	0.45	0.83		
Primary consolidation		C_C	1.15	0.95	1	1	0.45	0.95		
Secondary consolidation Tarao		$C_{\alpha\tau}$	0.141	0.11	0.184	0.184	0.024	0.110		
Secondary consolidation		C_α	0.03	0.03	0.04	0.04	0.01	0.03		
Bulk weight	kN/m³	γ_s	13	14	12	12	16	14		
Cohesion	kPa	C	0	0	0	0	0	0		
Cohesion increase		Δ_{cu}/Δ'_o	0.2	0.2	0.2	0.2	0.2	0.2		
Internal friction angle	°	ϕ	20	20	20	20	20	20		
Effective stress	kPa	σ'_o	7.65	20.3	26.3	27.6	30.9	35.5		
Influence factor of surcharge	%	I	100	100	100	100	100	100		
Calibration coefficient		β	1	1	1	1	1	1		

Drainage Parameters

Parameter	unit		Layer 1	Layer 2	Layer 3	Layer 4	Layer 5	Layer 6	Layer 7	Layer 8
Drain spacing (square)	m	D	0.6	0.6	0.6	0.5	0.5	0.5		
Drain diameter (MCD)	m	d	0.05	0.05	0.05	0.05	0.05	0.05		
Coefficient of horizontally Drained consolidation	m²/yr	C_h	0.35	1.6	1.2	0.4	0.7	0.4		

Fig. 14 illustrates the settlements recorded by the different surface settlement plates over the vacuum area along the filling process and the behaviour of the surface 3 months after the end of the Vacuum process.

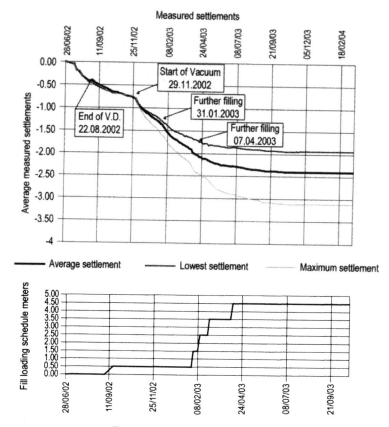

FIG. 14 *Settlement and fill record*

Even more interesting and not often available (Fig. 15), is the monitoring verified by the engineer after a period of 3 years and 10 months. Deformations include the added load of infrastructure construction and some live loads. Measurements were taken until 17th of March 2004 by Menard and from this period until today by the Engineer (IWB) in charge of the project.

11 CONCLUSION

The general concept of confinement of the soft muds by geotextile confined columns, the application of Vacuum consolidation under the future structural areas, the consolidation of the muds by vertical drains in non structural areas, have proven to be an adapted ground improvement solution for the "crash" program of the site preparation.

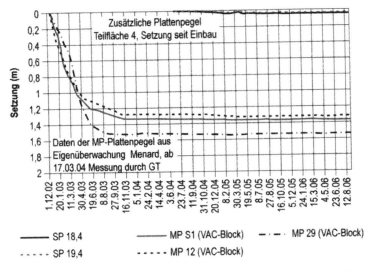

FIG. 15 *Longtime post construction and post construction settlement records (original document of the client) (x: dates of measurement; y: settlement in meters)*

A new application of Vaccum consolidation as retaining structure, proved to be successful.

The contractual utilization of a specially prepared spread sheet allowed to proceed with the works ahead of the finalization of the deformation criteria and schedule, avoiding argumentation on quantities and their financial consequences.

REFERENCES

Chu, J. (2009). State of the Art Report, Construction processes. 17th *ISCSMGG*, Alexandria, October 2009.

Chu, J. (2009). *Design and construction using preloading and vertical drain*, TC17 website: *http://www.bbri.be/go/tc17*.

Cognon, J. M. (1961). Vacuum consolidation. *Rev. French Geotechnique # 57*, (Oct.), 37-47

Cognon, J. M. (1961). Juran, I. and Thevanayagam, S. (1994). Vacuum consolidation technology – principles and field experience. *Proceedings of Settlement'94* – Sponsored by ASCE – Held June 16-18, 1994, College Station, Texas.

Faisal, A., Yee, K. and Varaksin, S. (1997). Treatment of Highly Compressible Soils. *Proceedings of the International Conference on Recent Advances in Soft Soil Engineering, Kuching*, Sarawak, March 1997.

Kirstein J. (2009). *Vacuum Consolidation* – TC17 website: *http://www.bbri.be/ go/tc17*.

Massé, F., Spaulding, C., Pr Ihm Chol Wong, Varaksin, S. (2001). *Vacuum consolidation: a review of 12 years of successful development.* Geo-Odyssey-ASCE/VIRGINIA TECH-Blacksburg, VA USA – June 9-13, 2001.

Marques, M. E. S., Leroueil, S., Varaksin, S. Demers, D., and D'Astous, J. (2000). *Vacuum Preloading of a Sensitive Champlain Sea Clay Deposit.* 53ª Conf. Canadienne de Géotechnique. Montréal, vol. (2): 1261-1268

Varaksin, S. (2003). Aménagement à Hambourg de la nouvelle usine AIRBUS 380 gagnée sur des vases hautement compressibles. *Salon et Congrès des Travaux Publics et du Génie Civil,* Paris.

Reiner, J., Stadie, R. (2002). *Geotechnishes Konzept der Flächenaufhörung im Mühlenberger Loch.* 27. Baugrundtagung der DGGT, 2002.

Raithel, M., Kempfert, H.-G. (1999). *bemessung von Goekunststoffummantelten Sandsäulen.* Bautechnik 76, Heft 11, Seiten 983-991, 1999.

Raithel, M., Kemfert, H.-G., Möbius, W., Wallins, P. (2002). *Gründungsmassnahmen zur Tragfähigkeitserhörhung und Setzungsreduktion beim Projekt Mühlenberger Loch* – Los 1. Geoteknic 25/Nr.1, Seiten 21-30, 2002.

Section 2 | Piled Embankments. Granular Piles and Deep Mixing

Keynote lecture

Deep mixing to improve the stability of embankments, levees, and floodwalls constructed on soft clay

Filz, G. M., Adams, T. E.
Department of Civil and Environmental Engineering, Virginia Tech, Blacksburg, USA

Navin, M. P.
U.S. Army Corps of Engineers, St. Louis, Missouri, USA

KEYWORDS deep mixing, numerical analyses, reliability analyses, stability, levees

ABSTRACT Deep mixing can be used to improve stability of embankments, levees and floodwalls on soft clay. However, design is complicated by the fact that these systems can experience multiple failure modes, such as composite shearing, column bending, column tilting, and racking failure associated with vertical displacement along weak joints in shear walls formed from overlapping columns. Furthermore, the strength of deep mixed ground is more variable than the strength of naturally occurring clay deposits. Numerical analyses and reliability analyses can be applied to address multiple failure modes and high strength variability so that economical and reliable designs can be developed. This paper presents examples of numerical and reliability analyses of embankments, levees, and floodwalls supported by columns and shear panels installed using the deep mixing method, including three examples of flood protection facilities in Louisiana for which the deep mixing method was applied after Hurricane Katrina.

INTRODUCTION

The purpose of this paper is to summarize key findings from a series of analyses performed to investigate the stability of embankments, levees, and floodwalls supported on deep mixed columns and elements in soft clay foundation soils. The paper includes sections presenting Background Information, Validation of Numerical Methods, Example Embankment Supported on Deep Mixed Columns, Reliability Analyses, and Three Levee and Floodwall Projects in Louisiana.

1 BACKGROUND INFORMATION

This section presents brief background information about the deep mixing method, typical geometry of deep mixing support systems, multiple failure modes that must be considered in design, and variability of the strength of deep mixed ground.

1.1 Deep Mixing Method

The deep mixing method blends cement, lime, slag, and/or other binders with soil to create columns, panels, or mass treatment zones that are stronger and less compressible than the *in situ* soil. The binder can be applied in dry form (the dry method of deep mixing) or in slurry form (the wet method of deep mixing). Typical stabilizer amounts range from about 100 to 500 kg of binder per cubic meter of soil to be treated. Single and multipleaxis rotary mixing tools are most common, with the binder being delivered through the hollow stems of the mixing tools. Other mixing methods, equipment, and innovations include pairs of horizontal rotating shafts with cutting and mixing blades to make barrettes, "chainsaw" type mixers to make continuous panels, rotating blenders attached to backhoes for mass stabilization, combinations of jetting with mechanical mixing, and rotating tools with adjustable mixing paddles to produce columns with variable diameters.

Additional background information about the deep mixing method is presented by Baker (2000), Broms (2003), Bruce (2000), Bruce and Bruce (2003), CDIT (2002), Elias et al. (2006), Esrig et al. (2003), EuroSoilStab (2002), Kitazume et al. (1997), Lambrechts (2005), Larsson (2005), Porbaha et al. (2001), Terashi (2003, 2005), Topolniki (2004), and others.

1.2 Geometry

Treatment depths typically range up to 15 m for the dry method and 30 m for the wet method, although much greater treatment depths are possible. Common column diameters range up to about 1 m for the dry method and up to about 2.5 m for the wet method.

Single-axis equipment can be used to create isolated columns, or the columns can be overlapped to create panels, cells, or blocks of treated ground. Multiple-axis equipment creates elements consisting of two or more overlapping columns in a single stroke, and these elements can be overlapped to form panels, cells, or blocks.

Deep mixed panels oriented perpendicular to the alignment of the embankment, levee, or floodwall can be effective for improving stability, particularly when placed under the side slopes of embankments or levees. Isolated columns can be used in the central area of embankments for settlement control. An embankment with isolated columns under the central portion and shear panels under the side slopes is shown in Fig. 1. Issues of load transfer and settlement of column-supported embankments are not addressed in this paper, but they are discussed by CDIT (2002), Chen et al. (2008), EuroSoilStab (2002), Filz (2009), Filz and Smith (2006), Lambrechts (2005), Lorenzo and Bergado (2003), and others.

1.3 Multiple Failure Modes

Designing stable support systems for embankments, levees, and floodwalls using the deep mixing method requires careful consideration of the various failure modes that can occur. Broms (1972) described several failure mechanisms, which are shown in Fig. 2, for piles used to stabilize slopes.

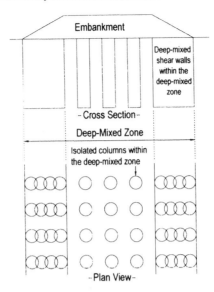

FIG. 1 *Schematic diagram of an embankment supported on deep mixed columns*

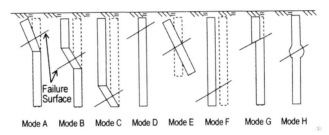

FIG. 2 *Potential failure mechanisms for deep mixed columns (after Kivelo and Broms, 1999)*

Failure modes A, B, and C in Fig. 2 represent column bending, D represents flow of soil around the column, E represents column tilting, F represents column translation, G represents shearing through the column, and H represents column crushing. Similar failure modes should be considered for deep mixed elements in foundation systems, as discussed by Filz (2009), Kivelo (1998), Kivelo and Broms (1999), Kitazume et al. (2000), Kitazume and Maruyama (2007a, b), CDIT (2002), Terashi (2005), and others. When overlapping columns are used to form shear panels as illustrated in Fig. 1, vertical shearing at the overlaps should also be considered, as discussed in Section 4.

1.4 Strength variability

The strength of deep mixed ground has relatively high variability. Statistical analyses were performed on 7,873 unconfined compression strength tests from 14 data sets for ten deep mixing projects in the USA, and the coefficient of variation ranged from 0.34 to 0.79, with an average value of 0.56 (Filz and Navin, 2006; Adams and Filz, 2007). Data from a collection of international projects show similar values (Larsson, 2005). For comparison, the coefficient of variation of undrained shear strength of naturally occurring clay deposits is typically in the range from 0.13 to 0.40 (Duncan, 2000). The relatively high variability of deep mixed ground can be addressed by performing reliability analyses, as discussed in Section 5.

2   VALIDATION OF NUMERICAL METHODS

Numerical stress-strain analyses have been used to evaluate stability of embankments, levees, and floodwalls founded on deep mixed columns (e.g., Han et al., 2005; Huang et al., 2006; Navin and Filz, 2006a, b; Adams et al. 2008a,b, 2009). In addition to shear failure, numerical analyses capture failure mechanisms not included in limit equilibrium slope stability analyses. Allowing other failure mechanisms, such as column bending and tilting, results in lower and more realistic values of factor of safety than when shearing is the only failure mechanism permitted.

Validation is necessary to have confidence in numerical analysis procedures. This section provides a summary of validation studies that have been performed using FLAC (Itasca, 2005) and FLAC3D (Itasca, 2002), and references are cited that provide the details. The validation studies involved comparing numerical analysis results with an instrumented case history, centrifuge model tests, and limit equilibrium analyses for comparable conditions.

A 5.5 m high test embankment supported on dry method columns at the I-95/Route 1 interchange in Alexandria, Virginia, provided a useful case history because the embankment was instrumented with slope inclinometer casings, pressure cells, and other instruments. Stewart et al. (2004) obtained good agreement between instrumentation data and calculations using FLAC for lateral deformations and vertical earth pressures.

While the I-95/Route 1 test embankment is a good source of data under service load conditions without failure, it would also be beneficial to validate numerical analysis procedures against failures. There are some instances of embankments supported on deep mixed columns that have failed, but it is difficult to obtain the data necessary to perform high quality analyses of the failures. Instead, published results of centrifuge model tests can be used to calibrate numerical methods under failure conditions. Validation analyses have been performed by Filz and Navin (2006) using the centrifuge model tests presented by Kitazume et al. (2000) and Inagaki et al. (2002), with good agreement obtained between observed and calculated failure modes and load versus deformation response.

To investigate the potential influence of three-dimensional effects, comparative analyses were performed by Navin et al. (2005) using FLAC and

FLAC3D for conditions of a model embankment in a centrifuge test performed by Inagaki et al. (2002). Additional discussions of two- and three-dimensional representations of embankments supported on deep mixed columns are presented by Huang et al. (2006)

The factor of safety values presented later in this paper for embankments, levees, and floodwalls were calculated using FLAC's automated factor of safety procedure, in which strengths are progressively reduced until a solution can no longer be obtained in a reasonable number of iterations. The inverse of the strength reduction factor is the factor of safety. To validate FLAC's automated factor of safety procedure, similar calculations were done manually for levees supported on deep mixed panels, and the same factor of safety values were obtained from automated and manual analyses. In addition, several comparative analyses were performed for soil slopes without deep mixed columns using FLAC's automated procedure and Spencer's method of limit equilibrium analyses, and the results are in good agreement. This work is described in a series of reports by Adams and Filz (2007, 2008, 2010).

3   EXAMPLE EMBANKMENT SUPPORTED ON DEEP MIXED COLUMNS

To assess the impact of multiple failure modes on stability of an embankment supported on deep mixed columns, the example embankment shown in Fig. 3 was analyzed by Filz and Navin (2006). The columns are relatively strong, and the factor of safety using limit equilibrium analyses is 4.4, which indicates a very stable configuration. However, the limit equilibrium analyses only permit a shearing mode of failure. Numerical stress-strain analyses, on the other hand, also allow the columns to bend and break. The factor of safety value from numerical analyses is 1.4, which is substantially lower than the value from limit equilibrium analyses. The shear strain contours and locations of tensile failure from the numerical analyses indicate a complex failure mechanism in which columns failed in bending and extension due to large deformations, and the soil between the columns experienced shear distortions as the broken parts of the columns bent and tilted.

The calculations for the example embankment shown in Fig. 3 were repeated, but with the deep mixed area under the embankment side slopes rearranged to form continuous panels of overlapping columns, as illustrated in Fig. 4. The same area replacement ratio of deep mixing is used in both cases, with the treated area rearranged to form continuous panels for the embankment in Fig. 4. Vertical joints were included at four locations in the shear walls, as shown in Fig. 4, to represent areas where columns overlap. The joints were assumed to have half the strength of the intact portions of the shear wall (CDIT, 2002; Broms, 2003). All other dimensions and material property values are the same for the cases represented in Figs. 3 and 4. Limit equilibrium analyses produced a factor of safety value of 4.4 for the conditions in

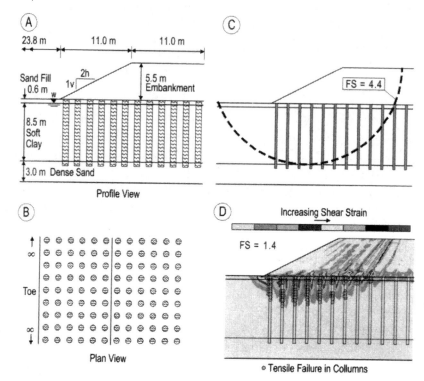

FIG. 3 *Example embankment supported on isolated columns: (A) profile view,
(B) plan view, (C) limit equilibrium factor of safety, and (D) numerical analyses
of factor of safety*

Fig. 4, which is the same value obtained for the embankment in Fig. 3. This
occurs because the area replacement ratio is the same for both cases and
limit equilibrium slope stability analyses only consider composite shearing
so they do not distinguish between the geometries shown in Figs. 3 and 4.
Numerical analyses, on the other hand, can distinguish between these cases
because the shear panels in Fig. 4 are not susceptible to the bending failure
mode that controlled performance in Fig. 3. The result is that the factor of
safety from numerical analyses for the shear panel arrangement in Fig. 4 is
3.1, which is much higher than the value of 1.4 for the isolated columns in
Fig. 3, even though the area replacement ratios are the same. The numerical
analyses allow racking distortions by sliding along the vertical joints in the
deep mixed zone, but the limit equilibrium analyses do not allow this failure
mode, and the factor of safety value of 3.1 from the numerical analyses is less
than the value of 4.4 from the limit equilibrium analyses.

For sufficiently low column strengths, limit equilibrium analyses may
give results similar to numerical analyses because composite shearing can
occur before the low strength columns bend and break. To determine if this

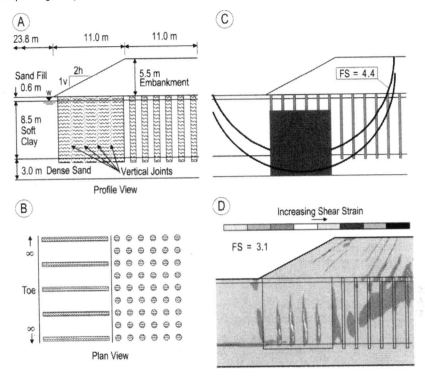

FIG. 4 *Example embankment supported on continuous shear panels under the embankment side slopes and isolated columns under the central portion of the embankment: (A) profile view, (B) plan view, (C) limit equilibrium factor of safety, and (D) numerical analyses of factor of safety*

occurs, several comparative limit equilibrium analyses and numerical stress-strain analyses were performed using an example embankment with isolated columns everywhere, similar to the example shown in Fig. 3. The ratio of factor of safety from limit equilibrium analyses to factor of safety from numerical analyses, FS_{LE}/FS_{NM}, is plotted versus the shear strength of the deep mixed columns in Fig. 5. Similar data from Han et al. (2005) are also included in Fig. 5 for analyses of another example embankment with different geometry and soil property values. For both sets of analyses, the ratio of FS_{LE} to FS_{NM} diverges from unity when the unconfined compressive strength of the deep mixed columns exceeds about 100 kPa (15 psi). These results show that limit equilibrium methods are suitable only for very low column strengths when isolated columns are used under the side slopes of embankments on soft clay.

4   RELIABILITY ANALYSES

Variability in the strength of deep mixed ground can be taken into account by performing reliability analyses. The example embankment in Section 4 was analyzed by Filz and Navin (2006), and the results are listed in Table 1.

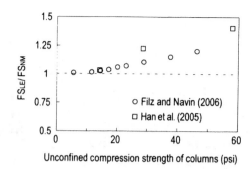

Unconfined compression strength of columns (psi)

FIG. 5 *Comparison of factors of safety from limit equilibrium and numerical methods for two example column-supported embankments (1 psi = 6.89 kPa)*

TABLE 1 RESULTS OF RELIABILITY ANALYSES FOR THE EXAMPLE EMBANKMENT

Case	Limit Equilibrium Method		Numerical Stress-Strain Method	
	FS	p(f)	FS	p(f)
Isolated Columns	4.4	0.01%	1.4	3.2%
Shear Panels	4.4	0.01%	3.1	0.01%

The probability-of-failure value from the numerical analyses is much higher than the value from the limit equilibrium analyses for the case with isolated columns everywhere, which is consistent with the lower factor of safety value for the numerical analysis due to the potential for bending and breaking of isolated columns. The limit equilibrium and numerical analysis results for continuous shear panels are much more similar, although the factor of safety value from numerical analyses is somewhat smaller than the value from limit equilibrium analyses because of the potential for shearing along vertical joints at column overlaps that was captured in the numerical analyses but not in the limit equilibrium analyses. For the numerical analyses of the embankment on isolated columns everywhere, it is also interesting to note that the factor of safety value of 1.4, which is in a normally accepted range for embankment design, corresponds to an unacceptably high value of 3.2% for the probability of failure. This occurs because of the high variability in the strength of deep mixed ground, as discussed in Section 2.4. Thus, safe design of embankments on deep mixed columns requires consideration of both multiple failure modes and high strength variability of deep mixed ground. The former can be captured by numerical analyses, and the latter can be addressed by reliability analyses.

5   LEVEE AND FLOODWALL PROJECTS IN LOUISIANA

This section presents numerical analysis results for three levee and floodwall projects in Louisiana showing the potential impacts of multiple failure modes on stability.

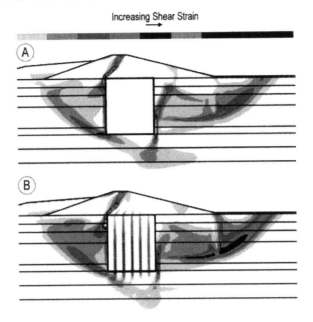

FIG. 6 *Shear strain contours for Plaquemines Parish Earth Levee: (A) 100% efficient vertical joints and (B) 0% efficient vertical joints*

5.1 Plaquemines Parish Earthen Levee

A 5.2 m high earthen levee was constructed on soft ground in Plaquemines Parish, Louisiana. The levee was supported and stabilized by shear panels formed of overlapping columns constructed by the dry method of deep mixing. The shear panels are 12.2 m deep, 10.7 m long, and positioned at a 2.1 m center-to-center spacing in the direction of the levee alignment. The columns are 0.8 m diameter, and the minimum specified overlap between adjacent columns was 0.15 m, which produces a chord length of 0.47 m at the overlap. Numerical analyses of stability were performed, and the results are shown in Fig. 6 for the condition of 100% efficient vertical joints in the shear panels, which is defined as the design overlap between adjacent columns, and 0% efficient vertical joints, which is defined as no overlap between adjacent columns. For 100% efficient vertical joints, the failure mode is rotation and translation of the deep mixed zone, and the factor of safety value is 1.51. For 0% efficient vertical joints, the failure mode includes shearing along the vertical joints in the deep mixed zone, and the factor of safety value is 1.37.

5.2 Gainard Woods Floodwall

An approximately 5 m high, reinforced concrete floodwall supported on steel piles was constructed in soft ground in Plaquemines Parish, Louisiana. To improve the stability of the floodwall, a zone of shear panels was installed on the protected side of the floodwall, between the floodwall and an existing drainage canal, to improve the stability of the system. The shear panels, which are oriented perpendicular to the floodwall alignment, were constructed using the deep mixing method by overlapping wet-mixed, triple-axis columns. The shear walls are 9.1 m long, 16.8 m deep, and positioned at a 1.83 m center-to-center spacing in the direction of the levee alignment. The columns are 0.91 m diameter, and the overlap between adjacent columns is 0.30 m, which produces a chord length of 0.68 m at the overlap. The vertical flood side limit of the DMM shear panels was located 9.1 m away from the protected side toe of the T-wall at the ground surface in order to provide a minimum of 3.0 m clearance between the bottom of the shear panels and the battered piles. This geometry leaves a gap between the DMM shear panels and the pile-supported T-wall, which created uncertainty about the effectiveness of the shear panels in improving stability of the T-wall.

The results of numerical analyses are shown in Fig. 7, which shows that the failure mode is complex, with high shear strains on the flood side of the flood-side piles, on the flood and protected sides of the deep mixed zone, and beneath the shallow slope on the protected side of the deep mixed zone. In addition, high shear strains occur in the deep mixed zone when the efficiency

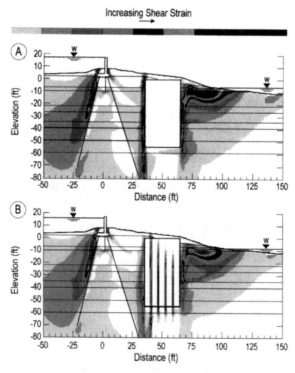

FIG. 7 *Shear strain contours for Gainard Woods Floodwall: (A) 100% efficient vertical joints and (B) 0% efficient vertical joints (1 ft = 0.305 m)*

of vertical joints is 0%. Although the shear strain diagrams look different, the factors of safety are similar at 1.59 for 100% efficient vertical joints and 1.57 for 0% efficient vertical joints. The small difference in the factor of safety for this case is attributed to the controlling influence of the shallow slope on the protected side of the deep mixed zone, which has a factor of safety equal to about 1.58 in limit equilibrium analyses.

5.3 Inner Harbor Navigation Channel

Stability analyses performed by the US Army Corps of Engineers indicated that an existing reinforced-concrete floodwall located in New Orleans Parish, Louisiana, had an inadequate factor of safety. The floodwall, which extends about 1.4 m above an adjacent soil berm on the protected side, is founded on a steel sheet-pile cutoff wall. The factor of safety was improved by installing shear walls consisting of overlapping columns installed by the dry method of deep mixing. The design called for shear walls about 9.8 to 11.3 m deep and 5.5 m long, with a diameter and spacing sufficient to produce an area replacement ratio of at least 30% and an overlap between adjacent columns equal to 20% of the area of an individual column.

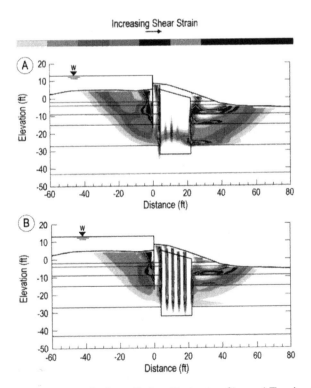

FIG. 8 *Shear strain contours for Inner Harbor Navigation Channel Floodwall: (A) 100% efficient vertical joints and (B) 0% efficient vertical joints (1 ft = 0.305 m)*

The results of numerical analyses are shown in Fig. 8, which indicates that the failure mode depends on the efficiency of vertical joints between adjacent columns. When the efficiency is 100%, the failure mode includes near horizontal shearing through the deep mixed zone, as well as shearing in the soil on both sides of the deep mixed zone. When the efficiency is 0%, the failure mode includes vertical shearing within the deep mixed zone, along with shearing in the adjacent soil similar to the 100% efficiency case. The factor of safety values are 1.38 and 1.15 for 100% and 0% efficient vertical joints, respectively.

6   CONCLUSIONS

The following conclusions can be drawn from the work presented and referenced above:

- The strength of deep mixed ground is more variable than the strength of typical soft clay deposits. This can produce a high probability of failure even when the factor of safety is in a customarily acceptable range.

- Multiple failure modes should be investigated when designing deep mixed support systems for embankments, levees, and floodwalls. Numerical analysis of stability can be an effective tool for this purpose, but such analyses should be validated against case history data, centrifuge model tests, and limit equilibrium analyses for comparable cases prior to using numerical analyses for design.

- Shear walls are more effective than isolated columns for improving stability of embankments, levees, and floodwalls.

- When shear walls are composed of overlapping columns, achieving continuity of the shear walls by effective overlaps can be important for stability.

ACKNOWLEDGEMENTS

The authors acknowledge the important contributions of Pete Cali, Mark Woodward, and Eddie Templeton. The information presented in this paper is based on work supported, in part, by the U.S. National Science Foundation under grant No. DGE-0504196. Any opinions, findings, conclusions, and recommendations in this paper are those of the authors and do not necessarily reflect the views of the U.S. National Science Foundation.

REFERENCES

Adams, T. E., Filz, G. M., Cali, P. R., and Woodward, M. L. (2008a). "Stability analyses of a levee on deep-mixed columns, Plaquemines Parish, Louisiana." *Geosustainability and Geohazard Mitigation*, GSP No. 178, ASCE, Reston, VA: 708-715.

Adams, T. E., Filz, G. M., Cali, P. R., and Woodward, M. L. (2008b). "Deformation and stability analyses of a pile supported t-wall with deep mixed shear panels in Plaquemines Parish, Louisiana." *Proc. 2nd Int. Conf. Geotechnical Engineering for Disaster Mitigation & Rehabilitation*, Science Press Beijing and Springer Berlin, 481-486.

Adams, T. E., Filz, G. M., and Navin, M. P. (2009). "Stability of embankments and levees on deep-mixed foundations." *Proc. Int. Symp. Deep Mixing & Admixture Stabilization*, Okinawa.

Adams, T. E., and Filz, G. M. (2007). "Technical memorandum: stability analyses of the P24 levee." *Report prepared for the New Orleans District, US Army Corps of Engineers*. 17 pages plus appendices.

Adams, T. E., and Filz, G. M. (2008). "Technical memorandum: stability analyses of the Gainard Woods Pump Station T-wall." *Report prepared for the New Orleans District, US Army Corps of Engineers*. 30 pages plus appendices.

Adams, T. E., and Filz, G. M. (2010). "Stability Analyses for the Inner Harbor Navigation Canal Reach III B-1A I-Wall." A report prepared for Burns Cooley Dennis, Inc., Blacksburg, Virginia.

Baker, S. (2000). "Deformation behaviour of lime/cement column stabilized clay." *Swedish Deep Stabilization Research Centre, Rapport 7*, Chalmers University of Technology.

Broms, B. B. (1972). "Stabilization of slopes with piles." *1st International Symposium on Landslide Control*, 115-123.

Broms, B. (2003). "Deep Soil Stabilization: Design and Construction of Lime and Lime/Cement Columns." Royal Institute of Technology, Stockholm.

Bruce, D. A. (2000). "An introduction to the deep soil mixing method as used in geotechnical applications." *FHWA-RD-99-138*, Federal Highway Administration, Washington.

Bruce, D. A., and Bruce, M. E. C. (2003). "The practitioner's guide to deep mixing." *Grouting and Ground Treatment*, GSP No. 120, ASCE, New Orleans, LA: 475-488.

CDIT (Coastal Development Institute of Technology). (2002). "The deep mixing method: principle, design, and construction." A. A. Balkema, Lisse, The Netherlands.

Chen, R. P., Chen, Y. M., Han, J., and Xu, Z. Z. (2008). "A theoretical solution for pile-supported embankments on soft soils under one-dimensional compression." *Canadian Geotechnical Journal.* 45: 611-623.

Duncan, J. M. (2000). "Factors of safety and reliability in geotechnical engineering." *J. Geotechnical and Geoenvironmental Engineering.* 126(4): 307-316.

Elias, V., Welsh, J., Warren, J., Lukas, R., Collin, J. G., and Berg, R. B. (2006). "Ground improvement methods-volume I." *Publication No. FHWA NHI-06-019*, Federal Highway Administration, Washington.

Esrig, M. I., Mac Kenna, P. E., and Forte, E. P. (2003). "Ground stabilization in the United States by the Scandinavian lime cement dry mix process." *Grouting and Ground Treatment*, GSP No 120, 501-514.

EuroSoilStab. (2002). "Development of design and construction methods to stabilise soft organic soils." Design Guide Soft Soil Stabilization, CT97-0351, Project No.: BE 96-3177.

Filz, G. M. (2009). "Design of Deep Mixing Support for Embankments and Levees." *Proc. Int. Symp. Deep Mixing & Admixture Stabilization*, Okinawa.

Filz, G. M., and Navin, M. P. (2006). "Stability of column-supported embankments," Virginia Transportation Research Council, Charlottesville, Virginia, 64 p.

Filz, G. M., and Smith, M. E. (2006). "Design of bridging layers in geosynthetic-reinforced, column-supported embankments." Virginia Transportation Research Council, Charlottesville, Virginia, 46 p.

Han, J., Parsons, R. L., Huang, J., and Sheth, A. R. (2005). "Factors of safety against deep-seated failure of embankments over deep mixed columns." *Proc. Int. Conf. Deep Mixing – Best Practice and Recent Advances, Deep Mixing'05*, SD Report 13 (CD-ROM), Swedish Geotechnical Institute, Linkoping, 231-236.

Huang, J., Han, J., and Porbaha, A. (2006). "Two and three-dimensional modeling of DM columns under embankments." *GeoCongress: Geotechnical Engineering in the Technology Age.*

Inagaki, M., Abe, T., Yamamoto, M., Nozu, M., Yanagawa, Y., and Li, L. (2002). "Behavior of cement deep mixing columns under road embankment." *Physical Modeling in Geotechnics: ICPMG'02*, 967-972.

Itasca Consulting Group, Inc. (2002). "FLAC3D Fast Lagrangian Analysis of Continua in 3 Dimensions." ITASCA Consulting Group, Minnesota, Minn., USA.

Itasca Consulting Group. (2005). *FLAC2D Fast Lagrangian Analysis of Continua, ITASCA Consulting Group*, Minneapolis.

Kitazume, M., Okano, K., and Miyajima, S. (2000). "Centrifuge model tests on failure envelope of column-type DMM-improved ground." *Soils and Foundations*. 40(4): 43-55.

Kitazume, M., and Maruyama, K. (2007a). "Centrifuge model tests on failure pattern of group column type deep mixing improved ground." *Proceedings of the Seventeenth International Offshore and Polar Engineering Conference*: 1293-1300.

Kitazume, M., and Maruyama, K. (2007b). "Internal stability of group column type deep mixing improved ground under embankment loading." *Soils and Foundations*. 47(3): 437-455.

Kitazume, M., Omine, K., Miyake, M., and Fujisawa, H. (1997). "JGS TC Report: Japanese design procedures and recent activities of DMM." *Grouting and Deep Mixing*, Yonekura, Terashi, and Shibazaki (eds), Balkema, Rotterdam, 925-930.

Kitazume, M., Okano, K., and Miyajima, S. (2000). "Centrifuge model tests on failure envelope of column-type DMM-improved ground." *Soils and Foundations*. 40(4): 43-55.

Kivelo, M. (1998). "Stabilization of embankments on soft soil with lime/cement columns." Doctoral Thesis, Royal Institute of Technology, Sweden.

Kivelo, M. and Broms, B. B. (1999). "Mechanical behaviour and shear resistance of lime/cement columns." *International Conference on Dry Mix Methods: Dry Mix Methods for Deep Soil Stabilization*, 193-200.

Lambrechts, J. R. (2005). "Design manual for deep mixing to support embankments over soft ground." US National Deep Mixing Program, Project 201.

Larsson, S. (2005). "State of practice report – execution, monitoring and quality control." *Proc. Int. Conf. on Deep Mixing Best Practices and Recent Advances*. (CD-ROM), Swedish Deep Stabilization Research Centre, Linkoping.

Lorenzo, G. A. and Bergado, D. T. (2003). "New consolidation equation for soil-cement pile improved ground." *Canadian Geotechnical Journal*. 40: 265-275.

Navin, M. P., Kim, M., and Filz, G. M. (2005). "Stability of embankments founded on deep-mixing-method columns: three-dimensional considerations." *Proc. 16th Int. Conf. Soil Mechanics and Geotechnical Engineering*, Osaka, (CD-ROM), Millpress, Rotterdam: 1227-1230.

Navin M. P., and Filz, G. M. (2006a). "Numerical stability analyses of embankments supported on deep mixed columns." *GeoShanghai 2006*, GSP No. 152, ASCE, 1-8.

Navin M. P., and Filz, G. M. (2006b). "Reliability of deep mixing method columns for embankment support." *Geotechnical Engineering in the*

Information Technology Age, Proc. of GeoCongress 2006. (CD-ROM), ASCE, Reston, VA: 6p.

Porbaha, A., Raybaut, J.-L., and Nicholson, P. (2001). "State of the art in construction aspects of deep mixing technology." *Ground Improvement.* 5(3): 123-140.

Stewart, M. E., Navin, M. P., and Filz, G. M. (2004). "Analysis of a Column-Supported Test Embankment at the I-95/Route 1 Interchange." *Proceedings of the GeoTrans 2004 Conference*, ASCE, 1337-1346.

Terashi, M. (2003). "The state of practice in deep mixing methods." *Grouting and Ground Improvement.* GSP No. 120: 25-49.

Terashi, M. (2005). "Keynote lecture: design of deep mixing in infrastructure applications." *Proc. International Conference on Deep Mixing- Best Practices and Recent Advances,* Stockholm.

Topolnicki, M. (2004). "In-situ soil mixing." Ch. 9 in *Ground Improvement, 2nd* Ed., M. P. Moseley and K. Kirsch, eds., Spon Press, New York.

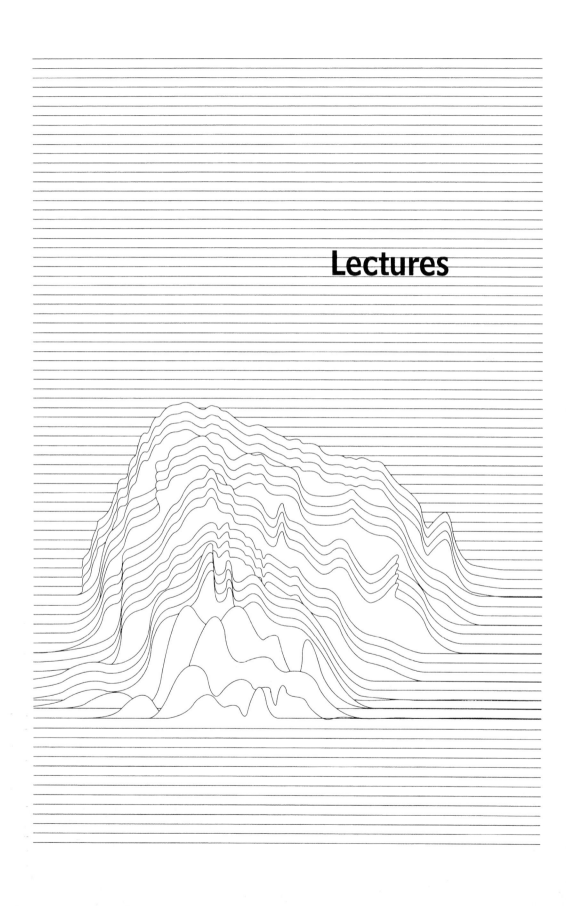

Lectures

Numerical Simulations and Full Scale Behavior of SDCM and DCM Piles on Soft Bangkok Clay

Bergado, D.T., Suksawat, T. and Saowapakpiboon, J.
Geotechnical and Geoenvirontal Engineering, School of Engineering and Technology, Asian Institute of Technology, Thailand

Voottipruex, P. and Jamsawang, P.
Department of Civil Engineering, King Mongkut's University of Technology North Bangkok, Thailand

KEYWORDS SDCM piles, DCM piles, bearing capacity, lateral load, settlement, lateral movement

ABSTRACT The new kind of reinforced Deep Cement Mixing (DCM) pile namely: Stiffened Deep Cement Mixing (SDCM) pile is introduced to mitigate the problems due to the low flexural resistance, quality control problems and unexpected failure of DCM pile. The SDCM pile consists of DCM pile reinforced with concrete core pile. Previously, the full scale pile load test and the full scale embankment loading test were successfully conducted in the field. To continue the study on the behavior of SDCM and DCM piles, the 3D finite element simulations using Plaxis 3D Foundation Software Version 1.6 were conducted in this study. The simulations of full scale pile load test consisted of two categories of testing, which are the axial compression and the lateral loading. For DCM C-1 and C-2 piles, the clay-cement cohesion, C_{DCM}, and clay-cement modulus, E_{DCM}, were obtained from simulations as 300 kPa and 200 kPa as well as 60,000 kPa and 40,000 kPa, respectively. For the SDCM piles, the simulation results show that increasing length ratio, L_{core}/L_{DCM}, increased the bearing capacity whereas the sectional area ratio, A_{core}/A_{DCM}, has only small effects on the bearing capacity for the axial compression loading. The verified parameters such as the clay-cement cohesion, C_{DCM}, and clay-cement modulus, E_{DCM}, from simulations of axial compression tests were 200 kPa and 30,000 kPa, respectively. On the other hand, increasing the sectional area ratio, A_{core}/A_{DCM}, significantly influenced the ultimate lateral resistance while the length ratio, L_{core}/L_{DCM}, is not significant in the ultimate lateral load capacity, when the length of concrete core pile is longer than 3.5 m. In addition, the tensile strength of DCM, T_{DCM}, and concrete core pile, T_{core}, are very important to the lateral pile resistance. The back-calculation results from simulations of tensile strength were 5,000 kPa and 50 kPa for the T_{core} and T_{DCM}, respectively. The simulations of the full scale embankment loading indicated similar results to the full scale vertical and lateral pile load tests. By increasing the length ratio, L_{core}/L_{DCM}, the surface and subsurface settlements reduced while the area ratio, A_{core}/A_{DCM}, only slightly reduced the settlements. In addition, the lateral movements of the embankment decreased by increasing both the length ratio, L_{core}/L_{DCM}, and the area ratio, A_{core}/A_{DCM}. Overall, the numerical simulations closely agreed with the field full scale test results and successfully verified the parameters affecting the performances and behavior of SDCM and DCM piles.

1   INTRODUCTION

Although Deep Cement Mixing (DCM) pile has many advantages with various applications, failure caused by pile failure can occur especially when subjected to the lateral loads. Moreover, the unexpected lower strength than the design commonly occurs due to lack of quality control during construction. Thus, DCM pile still fails by pile failure mode, which is lower than the soil failure mode particularly at the top of DCM pile due to low strength and stiffness as shown in Fig. 1 (a). In addition, Fig. 1 (b) shows the testing results of DCM pile on the soft Bangkok clay by Petchgate et al. (2003). About half of DCM piles failed by pile failure instead of soil failure. Consequently, the bearing capacity of DCM pile can be lower than the design load of 10 tons due to pile failure.

To mitigate the above-mentioned problem, a new kind of composite pile named Stiffened DCM (SDCM) pile is introduced. This composite pile is composed of an inner precast concrete pile hereinafter called concrete core pile and an external DCM pile socket, where the high strength concrete pile is designed to bear the load, and the DCM pile socket acts to transfer the axial force into the surrounding soil by skin friction.

The acceptance of numerical simulations in geotechnical problems is growing and finite element methods are increasingly used in the design of pile foundations. In this study, the full scale tests results were further simulated in order to study the parameters that affect the behavior of both the SDCM and DCM piles under the axial compression and lateral pile load as well as embankment load tests. Subsequently, the confirmed and verified parameters were used in the numerical experiments.

2   STIFFENED DEEP CEMENT MIXED (SDCM) PILES

Stiffened Deep Cement Mixed (SDCM) piles are a composite structure of concrete piles and deep cement mixed piles combining the advantages of both

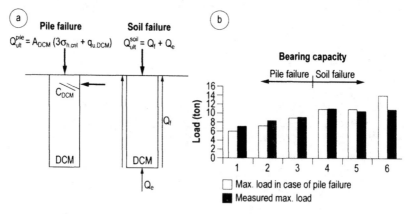

FIG. 1 *Low quality of DCM piles on soft Bangkok clay (Petchgate et al., 2003)*

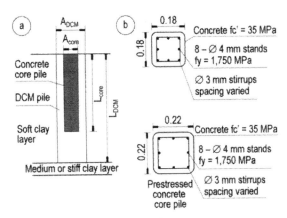

FIG. 2 *(a) Schematic of SDCM pile, (b) Details of prestressed concrete core piles*

components as shown in Fig. 2. A prestressed concrete stiffer core is installed by inserting into the center of a DCM pile immediately after the construction of wet mixing DCM pile. The two parts of the composite piles work together by supporting and transferring the vertical load effectively to the DCM pile and to the surrounding soil. In the SDCM pile, the DCM pile forms the surrounding outer layer supporting the concrete core pile increasing its stiffness and resisting compressive stress along the pile shaft. The dimensions of the two units should be such that both work together effectively and mobilize the full strength of the surrounding clayey soil. This novel method of improving the strength of DCM pile has been given different names by different researchers such as concrete cored DCM pile (Dong et al., 2004), composite DMM column (Zheng and Gu, 2005) and stiffened deep cement mixed (SDCM) column method (Wu et al., 2005).

2.1 DCM pile

The DCM pile is used as the socket pile in order to carry and transfer the load from concrete core pile to the surrounding soil. In this study, the DCM pile was constructed by wet mixing with 0.60 m diameter and 7.00 m length.

2.2 Concrete core pile

The Stiffened Deep Cement Mixing (SDCM) pile needs some material to enhance its stiffness like steel pile, timber and concrete pile etc. The prestressed concrete pile is more suitable than the other materials because it is cheaper than the steel pile and easier to manufacture. Moreover, the quality of prestressed concrete is better when comparing to the timber piles. Thus, the prestressed concrete core pile was proposed to bear compression in compression pile load test and resist lateral loads when subjected to the horizontal loads in SDCM pile.

2.3 Interface friction

The interface friction or adhesion means the ratio of the adhesive strength, τ_u, to the unconfined compression strength, q_u, of the clay cement. This value represents the frictional or adhesion resistance per unit cement soil strength provided on per unit side area of the concrete core pile. It is denoted by R_{inter} and calculated by following equation:

$$R_{inter} = \tau_u/q_u$$

where τ_u is the adhesive strength that can be calculated from the ultimate frictional strength P_u divided by the surface area of the stiffer core using the following equation:

$$\tau_u = P_u/AL$$

Many researchers have reported that the Rinter varies from 0.348 to 0.426 with an average value of 0.4 (Wu et al., 2005 and Bhandari, 2006).

3   PROJECT SITE AND SUBSOIL PROFILE

The full scale axial and lateral pile load tests were performed by Shinwuttiwong (2007) and Jamsawang (2008) and the full scale embankment load test was conducted by Jamsawang (2008) within the campus of Asian Institute of Technology (AIT). The site is situated in the central plains of Thailand, famous for its thick layer deposit of soft Bangkok clay. The foundation soils and their properties at the site are shown in Fig. 3. The uppermost 2.0 m thick layer is the weathered crust, which is underlain by 6.0 m thick soft to medium stiff

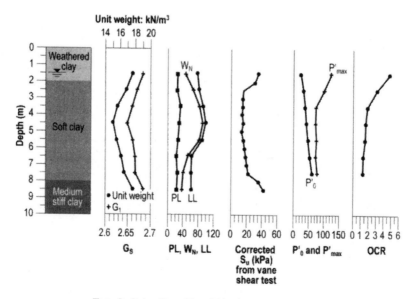

FIG. 3 *Subsoil profile within the campus of AIT*

clay layer. A stiff clay layer is found at the depth of 8.0 m from the surface. The undrained shear strength of the soft clay obtained from field vane test was 20 kPa and the strength of the stiff clay layer below the depth of 8.0 m from the surface is more than 40 kPa (Bergado et al., 1990). Other parameters are shown in Table 1.

The strength of the concrete piles was found to be 35 MPa. Two lengths of core piles were used in the field test namely: 4.0 m and 6.0 m. However, for the numerical simulation the length of the concrete pile was varied from 1.00 m to 7.00 m with 1.0 m increase to evaluate the effect of the lengths of the core pile on the capacity of the SDCM pile. The Mohr-Coulomb model was recommended to simulate for mass concrete core pile instead of linear elastic model because its stiffness can be overestimated if the tensile strain is large enough to crack the concrete (Tand and Vipulanandan, 2008).

4   FULL SCALE AXIAL AND LATERAL PILE LOAD AS WELL AS EMBANKMENT LOAD TESTS

The DCM pile was constructed by jet grouting method employing a jet pressure of 22 kPa and cement of 150 kg/m^3 of soil. The values of unconfined compressive strength of DCM obtained from field specimens ranged from 500 kPa to 1,500 kPa with average value of 900 kPa while the modulus of elasticity ranged from 50,000 kPa to 150,000 kPa with average value of 90,000 kPa indicating the empirical relation of $E_{50} = 100q_u$. The full scale pile load test piles consisted of 16 SDCM and 4 DCM piles. For the DCM pile 0.60 m in diameter and 7.00 m length was used and SDCM with lengths ranging from of 4 and 6 m was utilized.

4.1 Axial Compression Pile Load Test

The axial compression pile tests were conducted on both the DCM and SDCM piles. As shown in Fig. 2, the concrete core piles consisted of 0.18 m and 0.22 m. square piles. The DCM piles have 0.60 m in diameter. The load was applied increasing at 10 kN interval until pile failure. The bearing capacities of the 0.18 m square core pile with 4.00 m and 6.00 m were 265 kN and 300 kN, respectively, while the corresponding value for 0.22 m. square core pile with 4.00 m and 6.00 m were 275 kN and 315 kN, respectively. The bearing capacities of DCM piles, DCM-C1 and DCM-C2, were found to be 200 and 140 kN, respectively. The result from full scale pile load tests indicated that both the length and section area of concrete core piles increased the bearing capacities and reduced the settlement of SDCM piles. However, it was demonstrated that length was more dominant than the section area of the concrete core pile. Finally, the bearing capacity of SDCM pile is higher than the DCM pile.

Table 1: Soil models and parameters used in 3D FEM simulation

Materials	Depth (m)	Model	γ (kN/m³)	Material Behavior	E'_{ref} (kPa)	ν	λ^*	κ^*	c' (kPa)	ϕ' (deg)	OCR	Tensile Strenth (kPa)
Subsoil												
Weathered crust	0 - 2.0	MCM	17	Undrained	2500	0,25			10	23		
Soft clay	2.0 - 8.0	SSM	15	Undrained			0.10	0.020	2	23	1.5	
Medium stiff clay	8.0 - 10.0	MCM	18	Undrained	5,000	0.25			10	25		
Stiff clay	10.0 - 30.0	MCM	19	Undrained	9,000	0.25			30	26		
Foundation												
Concrete core pile		MCM	24	Drained	2.8×10^7	0.15			8,000	40		5,000
DCM pile (with interface elements)		MCM	15	Undrained	30,000 - 60,000	0.33			200 - 300	30		0 - 100
Concrete pile cap		LEM	-	Non-porous	2.1×10^7	0.15						

SSM: soft soil model; MCM: Mohr-Coulomb model; LEM: linear elastic model

4.2 Lateral Pile Load Test

The full scale lateral pile load tests were also conducted on designated SDCM piles. The 0.18 m and 0.22 m square core piles with 0.60 m DCM diameter were used. The horizontal load was applied at −0.30 m depth from the top of pile with increasing lateral load until pile failure. The maximum lateral load of the 0.18 m square core pile with length of 3.50 m and 5.50 m were 33 kN and 34.5 kN, respectively, while the maximum lateral load of the 0.22 m square core pile with length of 3.50 m and 5.50 m were 44.5 kN and 45.5 kN, respectively. By contrast, the maximum lateral load of DCM piles were only 3.5 and 2.5 kN for DCM L-1 and DCM L-2, respectively. The result indicated that the length of concrete core pile did not affect much the lateral capacity. However, the section area of the concrete core affects much the lateral capacity of the SDCM pile. Both the length and section area were significant reduction in pile displacement when the concrete core pile length was increased from 3.50 m to 5.50 m. Finally, the lateral bearing capacity of SDCM pile was found to be higher than the DCM pile.

4.3 Embankment load Test

Jamsawang (2008) and Jamsawang et al. (2008) constructed the full scale test embankment on improved soft Bangkok clay using two different methods namely: stiffened deep cement mixing (SDCM) pile and deep cement mixing (DCM) pile. The DCM pile consisted of 7 m long and 0.6 m in diameter. The objectives of this research work were to investigate ground improvement performances under embankment loading and to verify the related design parameters. Surface settlements and lateral movements were monitored during and after the embankment construction for two years. Fig. 4 shows the plan layout and side view of the embankment, respectively, together with the DCM and SDCM piles.

5 PROCEDURE OF SIMULATION

5.1 Procedure of numerical simulation of the axial compression and lateral pile load tests

Both axial compression pile load test and lateral pile load test were simulated by PLAXIS 3D Foundation software. The soft soil model (SSM) was used for the soft clay layer and the Mohr-Coulomb model (MCM) was used for other elements including DCM and SDCM piles. The soil modulus and parameters are tabulated in Table 1. Almost all of element used 15 node wedge element, except plate and interface elements that used the structural elements. The plate elements are based on the 8 node quadrilateral elements. The interface elements are different from the 8 node quadrilaterals that they have pairs of node instead of single node.

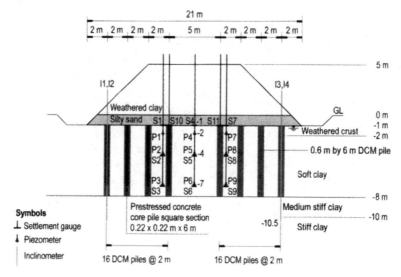

FIG. 4 *Side view and location of instrument of the embankment*

The initial stage was setup as the *in situ* state to generate the initial *in situ* stresses. The DCM pile and concrete core pile were then added to the simulation. The excavation stage was simulated by removing 1.00 m of soil around the pile for the axial compression pile load test and 1.5 m of the soil for the lateral pile load test. In the subsequent stages, a plate was used to distribute the load in the axial pile test and the pile cap was added to distribute the load in the lateral pile test.

After the addition of plate in case of axial load test and the pile cap in the lateral load case, the loading of the piles was commenced. For axial compression pile load test, the vertical load was increased in interval of 10 kPa until failure. For the lateral pile load test, the horizontal load was increased in interval of 5 kPa until failure. The programming of stage loading is illustrated in Fig. 5.

5.2 Procedure of numerical simulation of the full scale embankment load tests

The embankment is supported by two types of piles consisting of the 16-SDCM piles and 16-DCM piles. For the purpose of simulation, the length of concrete core piles in SDCM piles were varied from 3.00 to 7.00 m with varied sectional dimension from 0.22×0.22 to 0.30×0.30 m. The first phase was the initial stage that was setup as the *in situ* state (k_0 procedure) to generate the initial *in situ* stresses. In the second phase, the DCM pile and concrete core pile were installed. Next step was the excavation stage of the uppermost 1.00 m of soil. The subsequent steps consisted of filling the silty sand at the first phase at the base and subsequently, filled by weathered clay. Afterwards, the surface settlement at the top of SDCM, DCM and surrounding soil were checked after

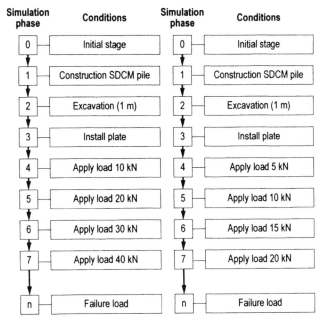

FIG. 5 *Finite Element simulation for axial compression and lateral pile load test*

60, 90, 120, 150, 180, 240, 300, 360, 420, 510 and 600 days, respectively. The details of the stage calculations are illustrated in Fig. 6.

6   RESULTS

6.1 Axial compression pile simulation

The appropriate parameters from back analysis for mixture of cement-clay cohesion in the DCM pile, C_{DCM}, obtained from the 3D finite element simulations were 300 kPa and 200 kPa, respectively, as illustrated in Fig. 7. However, the cement-clay modulus, E_{DCM}, were obtained as 60,000 kPa and 40,000 kPa for DCM C-1 and DCM C-2, respectively. Furthermore, for the SDCM pile, the corresponding value for C_{DCM} and E_{DCM} were 200 kPa and 30,000 kPa, respectively, as illustrated in Fig. 8. The slightly different results reflect the construction quality control in the field tests.

Fig. 9 shows the summary of the ultimate bearing capacity of SDCM pile which proportionally increased linearly with the increased lengths of concrete core pile while the sectional areas of the concrete core pile only slightly increased the bearing capacity. Consequently, increasing the length ratio, L_{core}/L_{DCM}, has dominant effect than increasing the sectional area ratio, A_{core}/A_{DCM}.

The mode of failure consisted of three categories, namely: concrete core pile failure, DCM pile failure, and soil failure. The SDCM pile failure occurred in the unreinforced part (DCM pile failure) because the DCM pile was not

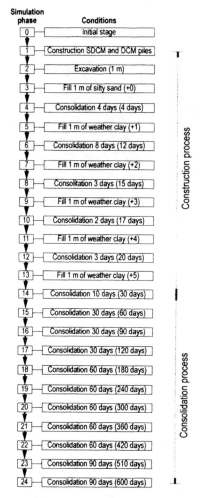

FIG. 6 *Finite element simulation for full scale embankment load test (accumulated time in parenthesis)*

strong enough to carry and transfer the load to the tip of DCM pile as demonstrated in Fig. 10.

6.2 Lateral load simulation

The appropriate values for mixture of cement-clay cohesion in the DCM pile, C_{DCM}, and mixture of cement-clay modulus, E_{DCM}, obtained from the 3D finite element simulation were similar to that one in the axial compression pile. In addition, the tensile strength of DCM pile, T_{DCM}, and tensile strength of concrete core, Tcore, were evaluated in this study. The T_{DCM} obtained from the simulation of DCM pile were 50 kPa and 25 kPa for DCM L-1 and DCM L-2, respectively, and the corresponding values for T_{core} and T_{DCM} obtained from the simulation were 5,000 kPa and 50 kPa, respectively (Figs. 11 and 12).

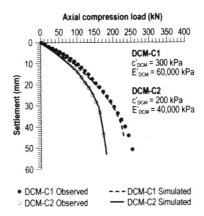

FIG. 7 *Comparisons between observed and simulated axial compression load – settlement curves for DCM-C1 and DCM-C2*

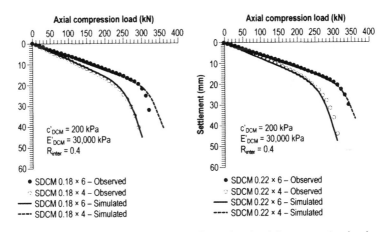

FIG. 8 *Comparisons between observed and simulated axial compression load – settlement curves for SDCM*

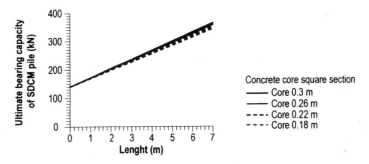

FIG. 9 *Effect of lengths and sectional areas of concrete core piles on ultimate bearing capacity*

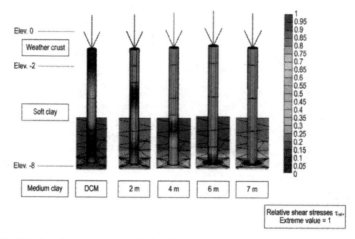

FIG. 10 *Relative shear stresses of 0.22 × 0.22 m core piles at failure load from simulations*

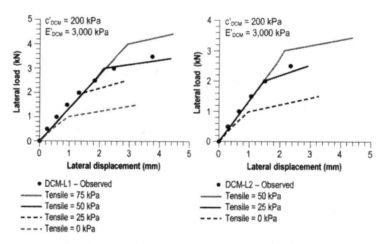

FIG. 11 *Comparisons between observed and simulated lateral load – settlement curves for DCM-L1 and DCM-L2*

The ultimate lateral load of SDCM pile increased with increasing sectional area because it increased the stiffness of the SDCM pile. However, the length of concrete core pile did not increase the ultimate lateral load capacity when using concrete core pile lengths longer than 3.5 m (Fig. 13).

6.3 Embankment load simulation

The surface settlements were measured at the top of DCM, SDCM piles and the unimproved ground in the middle of the embankment (untreated clay). The observed settlements are plotted in Fig. 14 together with the simulated values. Both the magnitude and rate of settlements from simulations agreed well with the observed data from field test as illustrated in Fig. 14. Consequently, the parameters involved were derived and verified. The parametric study

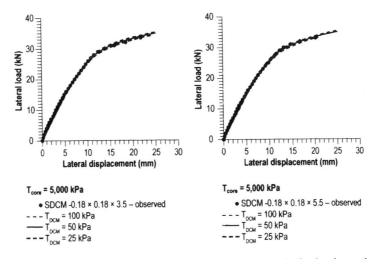

T_{core} = 5,000 kPa

● SDCM -0.18 × 0.18 × 3.5 – observed
- - - T_{DCM} = 100 kPa
—— T_{DCM} = 50 kPa
- - - T_{DCM} = 25 kPa

T_{core} = 5,000 kPa

● SDCM -0.18 × 0.18 × 5.5 – observed
- - - T_{DCM} = 100 kPa
—— T_{DCM} = 50 kPa
- - - T_{DCM} = 25 kPa

FIG. 12 *Comparisons between observed and simulated lateral piles load – settlement curves for SDCM*

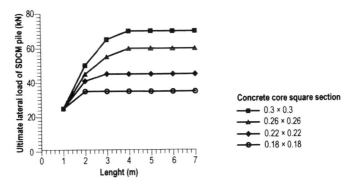

Concrete core square section
—■— 0.3 × 0.3
—▲— 0.26 × 0.26
—◆— 0.22 × 0.22
—⊘— 0.18 × 0.18

FIG. 13 *Effect of lengths and sectional areas of concrete core piles on the ultimate lateral load of SDCM pile*

was conducted by varying the sectional areas of the concrete core pile of 0.22×0.22 m and 0.30×0.30 m as well as varying the lengths of concrete core piles of 4, 5, 6 and 7 m to study their effects on the embankment settlements. The length ratio, L_{core}/L_{DCM}, significantly affected the settlements of the SDCM piles while the area ratio, A_{core}/A_{DCM}, has only small effects on the settlements of the SDCM piles. The effects of lengths and sectional areas of the concrete core piles of SDCM piles on the ultimate settlement of embankment simulation are illustrated in Fig. 15. It can be summarized that the ultimate settlement at 600 days after consolidation proportionally decreased with increasing lengths of concrete core piles from 4 to 6 m and only slightly decreased from lengths of 6 to 7 m. Moreover, the ultimate settlement only slightly decreased when increasing the sectional areas of the concrete core piles from 0.22 to 0.30 m.

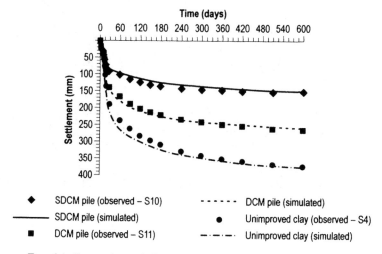

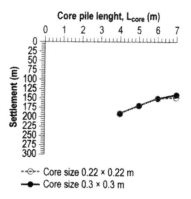

FIG. 14 *Comparison of observed and simulated surface settlements*

FIG. 15 *Effect of lengths and sectional areas of concrete core pile on ultimate settlements of SDCM pile*

Fig. 16 shows the summary of the effect of core pile length on the settlement at 600 days after the consolidation in surrounding clay of SDCM pile. The settlement of surrounding clay of SDCM at surface and 4 m depth linearly decreased with increasing the lengths of concrete core pile and only slightly decreased with increasing the sectional areas. Therefore, it can be concluded that the ultimate settlements proportionally reduced with increasing lengths of concrete core pile. In addition, both the sectional area and length of concrete core pile have no effect on the subsurface settlement at 7 m depth.

Differential settlements occur in the subsurface at various depths because the stresses proportionally decreased from the surface to the depths 4 and 7 m, respectively. Moreover, the stresses in the surrounding clay of SDCM and DCM piles as well as the unimproved zone are plotted together in Fig. 17. The stresses of surrounding clay of SDCM is the lowest meaning that the lowest

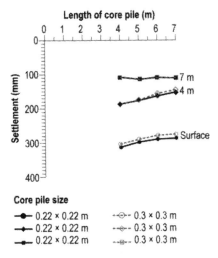

FIG. 16 *Effects of core pile lengths on ultimate surface and subsurface settlements in surrounding clay of SDCM pile*

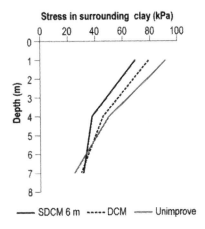

FIG. 17 *Stresses in surrounding clay of unimproved zone, SDCM and DCM piles piles at depth 1.4 and 7 m.*

settlements at the surface and 4 m depth. For the 7 m depth, the stresses are only slightly different, so the settlements were similar.

The effect of length of concrete core pile on the lateral movements are also studied through the simulations by varying the lengths of the concrete core pile from 4 to 7 m as well as their sectional areas consisting of 0.22 × 0.22 m and 0.30 × 0.30 m. The simulated and observed results of the lateral movements at 570 days after construction are illustrated in Figs. 18 and 19. The lateral movement significantly reduced with increasing lengths as well as with increasing the sectional areas of concrete core piles. It can be summarized

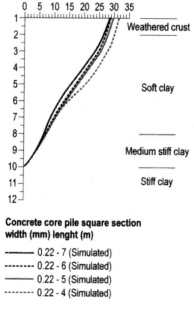

FIG. 18 *Effects of concrete core pile lengths on lateral movement profiles of SDCM pile with 0.22 × 0.22 m core pile from simulations*

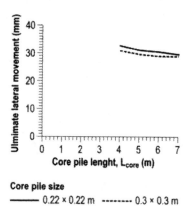

FIG. 19 *Effects of sectional areas and lengths of concrete core piles on the maximum lateral movement of SDCM pile*

that increasing both the lengths and sectional areas of core piles reduced the lateral movement.

7   CONCLUSIONS

7.1 Axial Compression Pile Test and Simulation

The appropriate parameter for mixture of cement-clay cohesion in the DCM pile, C_{DCM}, obtained from the 3D finite element simulations were 300 kPa and 200 kPa for DCM-C1 and DCM-C2, respectively. The mixture of cement-clay modulus in the DCM pile, E_{DCM}, were 60,000 kPa and 40,000 kPa for DCM-C1 and DCM-C2, respectively. For the SDCM pile, the corresponding value for C_{DCM} and E_{DCM} were 200 kPa and 30,000 kPa, respectively. The slightly different results reflect the construction quality control in the field tests.

The ultimate bearing capacity of SDCM pile proportionally increased linearly with the increased lengths of concrete core pile while the sectional areas of concrete core pile only slightly increased the bearing capacity when increasing the core pile sizes. Increasing the length ratio, L_{core}/L_{DCM}, has dominant effect to the axial capacity than increasing the sectional area ratio, A_{core}/A_{DCM}.

The axial load transfer along the pile length of the DCM and SDCM pile are different. For the DCM pile, the maximum load developed at the top 1 m and rapidly decreased until the depth of 4 m from the pile top at constant load of 10% of the ultimate load until the tip of DCM pile. Thus, the failure takes place at the top in the case of DCM pile. On the other hand, the axial load at the top of SDCM comprised 90% of ultimate load and linearly decreased to the pile tip consisting of 70% and 30% of ultimate load corresponding to 2 m and 7 m of concrete core pile length, respectively.

The mode of failure of SDCM piles consists of three categories, which are concrete core pile failure, DCM pile failure and soil failure. The SDCM pile failure occurred in the unreinforced part (DCM pile failure) because the DCM pile is not strong enough to transfer load to the tip of DCM pile.

7.2 Lateral Load Simulation

In the lateral load simulations, the appropriate value for cement-clay cohesion, C_{DCM}, and cement clay modulus, E_{DCM}, obtained from the 3D finite element simulation were similar to that one in the axial compression pile. In addition, the tensile strength of DCM pile, T_{DCM}, and tensile strength of concrete core pile, T_{core}, were evaluated in this study. The T_{DCM} obtained from the simulation of DCM pile were 50 kPa and 25 kPa for DCM L-1 and DCM L-2, respectively, which are similar to the back-calculated value of the tensile strength of the full scale pile load test (about 60 kPa and 30 kPa for DCM L-1 and DCM L-2). For the SDCM pile, the corresponding values for T_{core} and T_{DCM} obtained from the simulation were 5,000 kPa and 50 kPa, respectively. The ultimate lateral load of SDCM pile increased with increasing sectional area of the concrete core pile because it increased the stiffness of the pile. However, the length of concrete core pile did not increase the ultimate lateral load capacity when the lengths are longer than 3.5 m.

The maximum moments of DCM piles were 1.5 kN-m and 0.6 kN-m for the DCM L-1 and DCM L-2, respectively. The maximum moment was located about the excavation base similar to the observations in the field test. For the SDCM pile, the maximum moments depended on the concrete core pile sectional area where the bigger core size had higher moment. The magnitude of bending moments for 0.18 and 0.22 m section areas were 20 kN-m and 14 kN-m, respectively. These values were similar to the result of laboratory beam tests, which were 18 kN-m and 12 kN-m for concrete core pile of 0.22 m and 0.18 m section areas, respectively.

The failure mode of DCM pile occurred by the bending moment. The failure modes of SDCM pile can be divided into two categories, which are short and long piles. Both failure categories depended on the length and the sectional area of concrete core pile. For the SDCM pile with lengths longer than 3.5 m, the failure occurred by bending moment (long pile failure) while the short pile failed by the surrounding soil failure.

7.3 Embankment Load Simulation

The full scale embankment loading test supported by SDCM and DCM piles was constructed, monitored and consequently simulated using Plaxis Foundation 3D software in order to study and verify the design parameters. The appropriate parameter for cement-clay cohesion, C_{DCM}, and cement-clay modulus, E_{DCM}, obtained from the 3D finite element simulations were 200 kPa and 30,000 kPa, respectively. The result indicated that the longer concrete core pile can reduce the vertical displacement of SDCM pile as well as the subsurface portions of the surrounding soil. The settlement reduced linearly with increasing lengths of concrete core piles from 4 to 6 m but slightly reduced from 6 to 7 m core pile length. Moreover, the length of concrete core pile affected both the surface and subsurface settlements at 4 m but did not affect the subsurface settlement at 7 m.

In case of lateral deformation, the length and sectional areas of concrete core pile significantly affected the lateral movement of the embankment. The longer the lengths, the lower the lateral movements. Furthermore, the bigger sectional areas also reduced the lateral movements. It was also found that, the concrete core pile should be longer than 4 m in order to reduce the lateral movements of the embankment.

REFFERENCES

Bhandari, A. (2006). *Laboratory investigation of stiffened deep cement mixed (SDCM) pile*. M. Eng. Thesis No. GT-78-6, Asian Institute of Technology, Bangkok, Thailand.

Bergado, D. T., Ahmed, S., Sampaco, C. L. and Balasubramaniam, A. S. (1990). Settlement of Bangna-Bangpakong Highway on soft Bangkok clay. *ASCE Journal of Geotechnical Engineering*, Vol. 116, No. 1, pp. 136-154.

Brinkgreve, R. B. and Vermeer, P. A. (2006). *PLAXIS 3D Foundation Version 1.6 Manual*. Balkema, A. A., Rotterdam, Brookfield, Netherland.

Broms, B. B. (1964). Lateral resistance of piles on cohesive soils. , *Journal of The soil mechanic and foundation division, Proceedings of the American society of Civil Engineering,,* Vol. 90, No. SM 2, March, 1964.

Dong, P., Qin, R. and Chen, Z. (2004). Bearing capacity and settlement of concrete-cored DCM pile in soft ground. *Geotechnical and Geological Engineering,* Vol. 22, No. 1.

Jamsawang, P. (2008). *Full Scale Tests On Stiffened Deeep Cement Mixing (SDCM) Pile Including 3d Finite Element Simulation.* D. Eng. Diss. No. GE-08-01, Asian Institute of Technology, Bangkok, Thailand.

Jamsawang, P., Bergado, D. T., Bandari, A. and Voottipruex, P. (2008). Investigation and simulation of behavior of stiffened deep cement mixing (SDCM) piles. *International Journal of Geotechnical Engineering,* Vol. 2(3), pp.229-246.

Jamsawang, P., Bergado, D. T., Bandari, A. and Voottipruex, P. (2009). Behavior of stiffened deep cement mixing pile in laboratory. *Lowland International Journal.*

Jamsawang, P., Bergado, D. T. and Voottipruex, P. (2009). Laboratory and field behavior of   stiffened deep cement mixing (SDCM) pile. *Ground Improvement Journal,* (Accepted).

Jamsawang, P., Bergado, D. T. and Voottipruex, P. (2009). Behavior and 3D finite element simulation of stiffened deep cement mixing (SDCM) pile foundation under full scale loading. *Journal of Geotechnical and Geoenvironmental Engineering.*

Petchgate, K., Jongpradist, P. and Panmanajareonphol, S. (2003a). Field pile load test of soil-cement column in soft clay. *Proceedings of the International Symposium 2003 on Soil/Ground Improvement and Geosynthetics in Waste Containment and Erosion Control Applications, 2-3* December 2003, Asian Institute of Technology, Thailand, pp. 175-184.

Shinwuttiwong, W. (2007). *Full Scale Behavior of SDCM Piles under Axial and Lateral Loading with Simulations,* M. Eng. Thesis No GE-07-04, Asian Institute of Technology, Bangkok, Thailand.

Suksawat, T. (2008). *Numerical simulation of SDCM and DCM piles under axial and lateral loads and under embankment* load: *A parametric study.* M. Eng. Thesis No. GE-08-09, Asian Institute of Technology, Bangkok, Thailand.

Tand, K. E. and Vipulanandan, C. (2008). Comparison of computed vs. measured lateral load/deflection response of ACIP piles. *Plaxis Bulletin,* issue 23, pp. 10-13.

Wehnert, M. and Vermeer, P. A. (2004). Numerical analysis of load tests on bored piles. *NUMOT 9th,* Ottawa, Canada.

Wu, M., Zhao, X. and Dou, Y. M. (2005). Application of stiffened deep cement mixed column in ground improvement. *Proceedings International Conference on Deep Mixing Best Practices and Recent Advances,* Stockholm, Sweden.

Zheng, G. and Gu, X. L. (2005). Development and practice of composite DMM column in China. *Proceedings 16th International Conference on Soil Mechanics and Geotechnical Engineering*, Osaka, Japan, Vol. 3, pp.1295-1300.

Full-scale experiments of pile-supported earth platform under a concrete floor slab and an embankment

Briançon L.
Geotechnical department – Conservatoire national des arts et métiers – Paris, France

Simon B.
Terrasol – Montreuil, France

KEYWORDS piled embankment, concrete floor slab, load transfer, full-scale experiments
ABSTRACT This paper presents two full-scale experiments of pile supported earth platform under a concrete floor slab and an embankment. These experiments show that pile reinforcement combined with a load transfer platform leads to enhanced efficiency. Analytical and numerical approaches were developed to model the load transfer under the reinforced concrete floor slab. They demonstrate that it is possible to describe the mechanisms analytically and numerically if the models are calibrated from the results of the static vertical load test on a single pile. The results of the reinforced embankment show that the load transfer is the result of several mechanisms, depending on the complexity of the reinforcement. Measurements show that the load repartition above the granular platform depends on the type of its reinforcement. Some rules for evaluating the efficiency are proposed.

1 INTRODUCTION

Soft soil improvement by geosynthetics and rigid piles is common. However, some of the existing design methods do not take into account the complex behavior developed in these reinforced structures (Briançon et al., 2004). A French national research project (ASIRI) was thus launched to improve knowledge in this field and to draft a document constituting the guidelines for the design, construction and control of embankments and pavements on pile-reinforced soils. In this connection, two full-scale experiments of pile supported earth platform under a concrete floor slab and an embankment were carried out. From the important measurement data of these full-scale experiments, the mechanism developed in reinforced structures has been thoroughly examined.

2 FULL-SCALE EXPERIMENT UNDER CONCRETE SLAB

2.1 Site conditions and construction

The full-scale experimental sections are located in the industrial zone of Saint-Ouen l'Aumône, about 30 km Northwest of Paris, on the Oise River left bank.

Several boreholes and soil tests were carried out to characterize the subsoil profile; they indicated various soft soil layers (Figure 1). Dynamic and static cone penetration tests were driven to determine the depth of the bearing layer, which consists of a dense sand layer. These tests have highlighted variations of the soft soil thickness across the experiment site. The ground water table is approximately 2–3 m below the ground surface. A typical cross-section of the soft soil, piles, granular platform, floor slab and embankment used as fill load is shown schematically in Fig. 1.

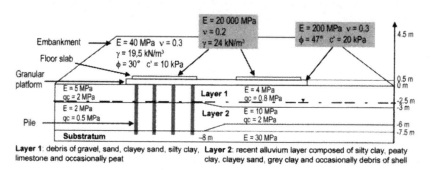

Layer 1: debris of gravel, sand, clayey sand, silty clay,   **Layer 2**: recent alluvium layer composed of silty clay, peaty
limestone and occasionally peat   clay, clayey sand, grey clay and occasionally debris of shell

FIG. 1 *Schematic cross section and geotechnical profile*

Two full-scale concrete floor slabs were built, one on a non-reinforced section to serve as reference and the other on a reinforced section. The latter section, with a 10 m by 10 m area, was reinforced by 4 by 4 concrete piles, 0.42 m in diameter, set up along a 2.5 m square grid. Piles were installed with a non-displacement technique. The replacement ratio of the reinforced section was $\alpha = 2.2$ %. The piles were embedded 0.5 m into the substratum and the pile heads were struck off to ground level.

The load transfer platform, with a total 0.5 m thickness, was formed by compacted granular fill ($\phi' = 46.9°$ and c' = 50 kPa). Bottom 0.25 m fill layer was prepared before pile installation to serve as traffic layer for the pile equipment. Floor slabs which had 8 m by 8 m square area and 0.17 m thickness were made of steel fibers reinforced concrete. Both sections were loaded in two stages, namely with 1.5 m backfill left for three months (July - October 2006), followed by 2.5 m additional backfill.

A static vertical load test was carried out on a single floating pile. The pile failed under an axial load equal to 400 kN, 74 % of this load being equilibrated by shaft friction and 26 % by end-bearing.

2.2 Instrumentation

Instrumentation of the non-reinforced section comprised a horizontal inclinometer at the base of granular platform along one of its diagonals. In the reinforced section, differential settlement between pile heads and compressible soil was also measured with a horizontal inclinometer. The load transferred to pile heads was measured by earth pressure cells (EPC) of the same size area as piles (Fig. 2).

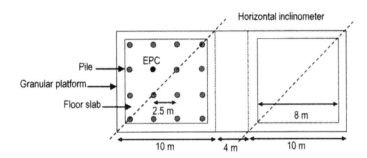

FIG. 2 *Layout of the experimental sections, instrumentation and piles*

2.3 Result and analysis

2.3.1 Load transfer

The total soil pressure measured on the pile head shows that the stress applied on the pile head increases quickly with fill embankment loading stage imposed and reaches 590 kPa after the first stage and 1800 kPa after the second stage. The recorded total soil pressure acting on the pile head is always higher than the static overburden pressure, which proves that the soil stress distribution towards the pile heads occurred as soon as the fill embankment was loaded on the reinforced section.

2.3.2 Settlement

The total settlement measured in non-reinforced section under the slab reaches 25 mm after the first stage and 70 mm few months after the second stage.

Fig. 3 shows the profile of differential settlement between a pile and the soil in reinforced section for few dates. It is presented for a half diagonally of grid in the middle of the section. A half diameter pile illustrates the position of the pile compared to the point of differential settlement. We observe that the shape of the settlement is very flat from 0.8 m of the pile circumference. The differential settlement reaches 23 mm one year after the second stage loading.

2.3.3 Analytical and numerical approaches

Analytical and numerical approaches were developed to model the load transfer in reinforced structures.

The analytical approach considers an elementary reinforced cell decomposed in two subdomains: the pile and its soil cylindrical vertical extensions

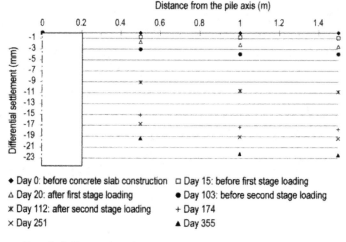

FIG. 3 *Differential settlement measured in a reinforced section*

and the all remaining soil volume (Cuira and Simon, 2009). It is assumed that the interaction between both domains can be fully described by shear stress τ developed on their common boundary and built in reference to t-z curves commonly used in pile foundation design. In France, these t-z curves are based on Menard pressuremeter modulus E_M and values of the limit unit shaft friction q_s and end bearing q_p (Frank and Zhao, 1982). The analytical approach has been introduced in a specific module "Taspie+" of the oriented foundation design software "Foxta v3" developed by Terrasol.

The numerical approach consists in an axisymmetrical model of the elementary cell (Plaxis). Constitutive model for all soil layers is the linear-elastic-perfectly-plastic Mohr Coulomb. Elastic model is adopted for the floor slab and pile. Interfaces are introduced between soil and pile with a Mohr-Coulomb failure criterion and the following properties: the constrained modulus is taken the same as the surrounding soil; the friction angle is null and the cohesion is equal to the ultimate skin friction appropriate for that layer. The soft soil strength parameters are determined from the triaxial tests made on selected undisturbed samples. The substratum parameters have been adjusted to give the same load curve at the pile tip as the one observed during the loading test on a single pile carried out in our experiment.

From the soil model considered, the settlement of the non-reinforced section has been calculated analytically: the result (= 77 mm), in agreement with the measured settlement, validates the choice of the soil model.

The numerical axisymmetrical model compared with the analytical model for a case with a single loaded pile in a 2.5 m square grid validated the calibration of the numerical model parameters.

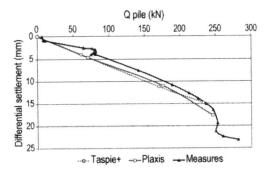

FIG. 4 *Differential settlements measured and calculated in the reinforced section*

Both approaches were used to simulate the behavior of the reinforced concrete floor slab. Figure 4 shows the comparison of the load applied on pile versus the differential settlement for the measurement, numerical and analytical designs. A good agreement between the three curves is observed although both approaches do not take into account the discharge of the pile after the first loading. The profiles of differential settlements between piles and soil in reinforced section calculated with numerical and analytical approaches are similar to those measured. The concentration of stress on pile calculated by both approaches is the same that measured (Fig. 5a). The load transfer calculated at the pile level and under the concrete slab (Fig. 5) shows that there is a stress concentration in the cylindrical soil between the pile and the concrete floor slab and that the load transfer does not occur by arching (Fig. 5).

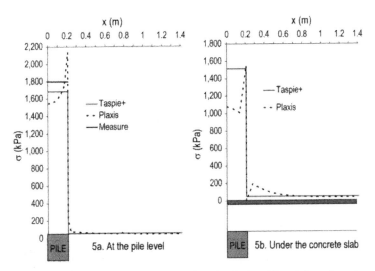

FIG. 5 *Stress distributions measured and calculated in reinforced section*

3   FULL-SCALE EXPERIMENT UNDER EMBANKMENT

3.1   Site conditions and construction

The site, located 20 km Northeast of Paris, France, was made available by the "Conseil Général de Seine & Marne". The thickness of the soft soil layer was found to be between 8 m and 10.5 m (Fig. 6). The soft soil consisted of clay and sandy clay of low plasticity and medium compressibility. A preexisting clayey fill, less than 2 m thick, covered the soft soil. The ground water level was found to be 2 m deep. The selected experiment test area is divided into four instrumented sections (1R, 2R, 3R and 4R). Rigid piles, made with soil displacement, reinforced three of the sections (2R, 3R and 4R). For all reinforced sections, each pile, 0.38 m in diameter, has its tip slightly embedded in the bearing layer (0.3 m) and its head cut off level with the traffic layer base. Another non-reinforced section (1R) is included for reference. The embankment is 5 m high.

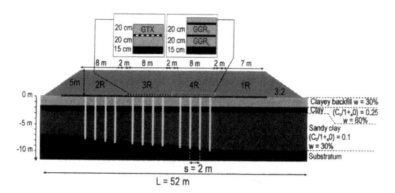

FIG. 6 *Typical site cross-section and geometrical characteristics*

Fig. 6 shows the typical cross section with soft soil, piles, load transfer platform and embankment. Each reinforced section has an 8 m by 8 m square area under full height embankment and it is reinforced by 4 by 4 piles defining a 2 m square grid. A load transfer platform made by compacted granular fill is laid at the fill base of sections 3R and 4R. One geotextile layer and two geogrids for sections 3R and 4R respectively enable the reinforcement of the granular platform.

A static load test was carried out on an embedded single pile. The embedded pile failed under an axial load equal to 600 kN, 39% of which being equilibrated by shaft friction and 61% by end-bearing.

3.2   Instrumentation

More than 70 sensors were installed in the load transfer platform, soft soil and concrete piles. In particular, load transfer was measured with earth and concrete pressure cells (E) and settlement with pressure transmitters for level measurement (T). The geosynthetic strain was monitored with optical fibres using the technique of Bragg gratings and called Geodetect. Fig. 7 shows the localization of sensors in section 3R, which is the same in sections 2R and 4R.

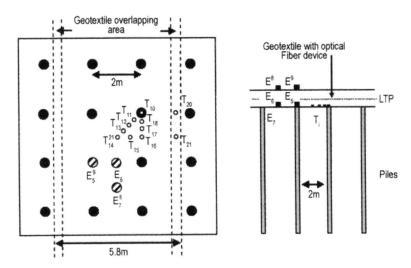

FIG. 7 *Instrumentation of section 3R under the embankment*

3.3 Result

3.3.1 Load transfer

In section 2R, the stress measured at the pile head (σ_{E1} = 570 kPa) is quite lower than that corresponding to a complete load transfer (estimated to 3,350 kPa). In sections 3R and 4R, the stress measured at the pile head shows that for both sections, the reinforcement system transfers a much more significant load towards the pile heads (Fig. 8). On the top of the load transfer platform, the stress above the pile in section 3R (σ_{E9} = 500 kPa) differs from that in section 4R (σ_{E14} = 130 kPa), although stresses on pile heads are very similar at the end of the embankment construction. This shows that mechanisms in the load transfer platform depend on its reinforcement.

3.3.2 Settlement

Although the compressibility of soft soil in the experimental area was not high (the settlement of non-reinforced embankment is only equal to 0.26 m), the settlement has been reduced through reinforcement by piles and reinforced granular platform (Table 1).

In reinforced sections, the profiles of settlement become very flat, more than 0.2 m from the pile shaft. Differential settlement begins during the embankment construction. For both sections with a load transfer platform, the

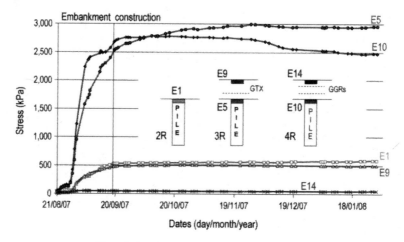

FIG. 8 *Load transfer in reinforced sections*

TABLE 1 SETTLEMENT MEASURED IN SECTIONS

Section	Settlement of soil (mm)	Differential settlement soil/pile (mm)
1R	260	—
2R	105	97
3R	71	41
4R	64	37

differential settlement remains constant after the end of embankment construction, despite a slight increase in the settlement of piles caused by high stress at the head as well as limited embedment at the toe.

3.4 Analysis

3.4.1 Pile behavior

Comparison between the curves of the pile head settlement versus the pile head stress and the curve obtained from the static load test on an embedded single pile shows that reinforcement piles of a full-scale experiment mostly mobilize their end-bearing capacity. This reflects both the group effect between all reinforcement piles and the influence of shallow embedment in the bearing layer. In the present case, the shallow pile embedment enables us to reduce the differential settlement between soft soil and piles at the pile head height by allowing piles to drive in. This parameter must clearly be taken into account by design to obtain a stable equilibrium at the end of the embankment construction together with the smallest differential settlement from then onwards.

3.4.2 Load transfer in granular platform

The reported experiment highlights the reduction of the stress applied on soft soil for sections with a load transfer platform (Fig. 9). For these sections, soft soil, which is increasingly loaded during the embankment construction, reveals a slight unloading from then onwards. For the section without a load transfer platform, the stress continues to increase after the construction.

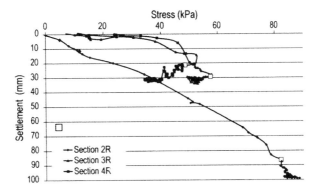

FIG. 9 *Settlement versus stress measured on soft soil*

Measurements of shear strain confirm that the mechanisms in the load transfer platform depend on the geosynthetics used: for section 3R, the tensile force of the geotextile layer increases during the embankment construction, especially near the pile heads, for section 4R, the geogrids start being stretched from the moment when the granular platform is compacted.

In addition, while the pile head stress was decreasing in section 4R (Fig. 8 - E_{10}), the stress measured just above the load transfer platform (Fig. 8 - E_{14}) remained constant. We can thus conclude that the load transfer mechanisms develop mainly in the load transfer layer itself and around the pile heads, and hardly modify stresses acting on the top of the load transfer layer, near the pile centre line.

4   DISCUSSION ABOUT THE EFFICIENCY

Evaluation of the efficiency of piled embankments was proposed mostly in reference to laboratory tests without any consideration of the mechanisms that could develop beneath the load transfer platform. Attention was restricted to the total vertical stress applied by embankment, the vertical stress acting on top of the columns, and the average vertical stress acting on the soil surface in between the columns.

Table 2 presents the reduction of stress applied on soft soil and the reduction of settlement of soft soil for the three reinforced sections comparatively to the non-reinforced section. It shows that a settlement reduction of 60% requires a stress reduction of only 13% for section 2R. This difference between the reduction of stress and the reduction of settlement is also observed in sections 3R and 4R. This confirms that it would be more relevant to define the efficiency of piled embankments with settlement parameter rather than with stress parameter.

TABLE 2 STRESS REDUCTION ON SOIL AND SETTLEMENT REDUCTION OF SOIL

Section	Settlement reduction (%)	Stress reduction (%)
2R	60	13
3R	73	50
4R	75	70

5   CONCLUSIONS

A French national research project (A.S.I.RI.) was launched to improve the knowledge in the field of pile supported earth platform and to draft a document constituting the Guidelines for the set up and the design of embankments and pavements on ground reinforced by rigid piles. Regarding this aim, two full-scale experiments were carried out: under a concrete floor slab and under an embankment. Measurement highlights the load transfer is different in granular platform under an embankment from that under a concrete floor slab: there is a stress concentration in the cylinder of soil between the pile and the concrete floor slab, the load transfer takes place with inverted pyramidal shape below the embakment and depends on the geosynthetic used for the reinforcement. These mechanisms could be modeled with great precaution in the calibration of the models with the modeling of the loading test on a single pile if it has been carried out on the site. Lastly, it seems to be more relevant to define the efficiency of this type of reinforcement with settlement criteria.

ACKNOWLEDGEMENTS

The authors would like to thank the French national project (A.S.I.RI.) for funding this research within the partnership between Fondasol, IREX, Keller, Ménard, Twintec LCPC, EGIS, Rincent BTP, Socotec, Tencate Geosynthetics. This work was made possible thanks to the financial support of Drast and RGCU and the "Conseil Général de Seine et Marne" and the "Port Autonome de Paris" who kindly allowed us to use the experimental site.

REFERENCES

Briançon, L., Kastner, R., Simon, B. and Dias, D. (2004). Etat des connaissances - Amélioration des sols par inclusions rigides. *Proc. Int. Symp. on Ground Improvement (ASEP-GI 2004)*, 9-10 September 2004, Presses de l'ENPC, 15-43 (In French).

Cuira, F. and Simon, B. (2009). Two simple tools for evaluating the complex interactions in a soil reinforced by rigid inclusions. Proc. 17th Int. Conf. on Soil Mechanics & Geotechnical Engineering, Alexandria, Egypt 5-9 October 2009, 1663-1666.

Frank, R. and Zhao, S. R. (1982). Estimation par les paramètres pressiomètriques de l'enfoncement sous charge axiale des pieux forés des sols fins. Bulletin de liaison des laboratoires des Ponts et Chaussées, 119, 17-24 (In French).

Piled embankments in the Netherlands; how to decide, how to design?

Eekelen, Suzanne J. M. van
Deltares, Technical University Delft, chair person Dutch CUR Task group 'Design Guideline Piled Embankments'
P.O. Box 177, 2600 MH Delft, The Netherlands, Suzanne.vanEekelen@Deltares.nl

Venmans, Arjan A. M.
Deltares and Technical University Delft

KEYWORDS Piled embankments, decision support systems, road construction methods, soft soils, design guideline, history

ABSTRACT In the 1930s, public opinion and minimizing traffic hindrance were carefully considered when making choices for road construction methods in the Netherlands. Later, technical and financial considerations became more leading and the public opinion less. Now, decision processes are changing further and criteria as whole life costs and traffic hindrance are becoming more important again. In addition, new construction methods, such as piled embankments, including their design guidelines have become available.

The last years have brought decision support systems to quantify the impacts of choices, making the selection process more transparent. A case study for the widening of the highway A2, and in addition several successful examples show that a contractor is stimulated to offer innovative solutions such as piled embankments when the principle sets requirements for construction time, protection of vulnerable objects or a period of maintenance for the contractor.

When design guidelines for these innovative solutions become available, like the recently introduced Dutch Piled embankment Design Guideline, the application of these techniques is further stimulated and the public demands are more easily fulfilled.

1 THE HISTORY OF ROAD CONSTRUCTION IN THE NETHERLANDS

During the last eight centuries, the Dutch have built nearly half of their country themselves: the polders. Usually, the weak and compressible soil in these polders gives major problems while constructing roads or railroads. The subsoil typically consists of 6 to 18 m of very soft and compressible organic clay and peat deposits, and below that a firm stratum of sand. The ground water is table just below ground surface. During many centuries, the Dutch in these areas therefore mainly chose to travel by boat.

However, cars and trains were developed and the need for road and railroads grew. So, the Dutch were faced with the problem of raising sand embankments on ground that should be considered unsuitable. A project that finally succeeded was the construction of the railroad between Gouda and Schoonhoven, see Fig. 1.

FIG. 1 *Railroad Gouda-Schoonhoven; construction 1855-1914, in service 1914-1942*

It took 59 years, from 1855 to 1914, to construct the around 16 km long railway. Five times the scheduled amount of sand was needed to finish it, which made 65 m^3 sand per meter railroad. Finally, the trains have only been in service during 28 years until 1942, when the occupier removed the rail steel and melted it down for war purposes.

FIG. 2 *Need for wider and more reliable roads in the thirties of the last century, source: www3.picturepush.com/photo/a/2223576/1024/Friese-B-nummers-trucks/12515.jpg*

Since the 1930s, the need for more reliable and wider roads (Fig. 2) increased with the development of more reliable diesel engines.

Keverling Buisman and his co-workers at the Dutch Institute for Soil Mechanics developed several construction methods to build roads in the soft Dutch polders (Heemstra, 2008). They for instance developed the road on a lightweight embankment (dried peat), the road on piles (concrete slabs on timber piles) and embankments reinforced with fascines (Fig. 3). In order to prevent rotting of the fascines, these mattresses needed to remain completely below ground water table during their service life.

In 1937, Keverling Buisman considered both applying a fascine mattress and a road on piles for the regional road N210 through the Krimpenerwaard polder. He finally recommended constructing a piled road to prevent high maintenance costs, and to "prevent complaints and unfavourable comments"

FIG. 3 *Fascine mattress in construction*

because of traffic hindrance. He thought that the piled road would meet more "appreciation". In other words, the public opinion was important if not a leading aspect for Keverling Buisman.

Between 1950 and 1980 the Dutch highway network grew from approximately 500 km to 3000 km. Construction of embankments on soft soil became common practice, and a standard practice developed. There was a gradual development in soil improvement from excavating and filling cunettes, via installation of sand drains to installation of prefabricated drains. Construction took years. Contracts were awarded on the basis of the lowest price for a design prepared by the principal. Decisions were motivated by technical considerations, not by the opinion of the general public as in the days of Keverling Buisman.

Times changed. New materials became available. For example, the development of geosynthetic reinforcement material made it possible to combine fascine mattresses and piled roads with concrete slabs into piled reinforced embankments. Newly developed guidelines give more security and safety for these new construction types.

Times changed for the decision processes as well. More people and more stakeholders have become involved. Road users cry out for solutions for the growing traffic jams. Criteria such as whole life costs and traffic hindrance are becoming more important again. Decision support systems have been developed to quantify and keep track of the impacts of design choices. Keverling Buisman would have welcomed this development!

This paper describes the current state-of-the-art in the Netherlands with respect to selection of construction methods for embankments on soft soil, with emphasis on the considerations related to piled embankments. In the last decade, around 23 piled embankments have actually been constructed in the Netherlands. Why were these actually built? In the beginning of 2010, the Dutch Guideline for the design of piled embankments was introduced. The clear design rules in this guideline give clarity, which makes decision processes easier.

2   PILED EMBANKMENTS IN THE NETHERLANDS

In 1999, the construction of the first Dutch piled embankment started. Since then, the Dutch have constructed around 22 piled embankments for roads. In 2008 the first railroad was build on a piled embankment (Van Duijnen and Van Eekelen, 2010). The total area of these 23 piled embankments is around 300,000 m^2.

From the 23 piled embankments, eight have been constructed for local authorities (towns), six for the Dutch ministry of Public Works, five for regional authorities (provinces) and the others for the Dutch Railways and a knowledge institute. One of the most remarkable piled embankments is the new N210 in the Krimpenerwaard polder, which is a 14 km long regional road (Haring et al., 2008). The next chapter describes the main reasons for the choice for this, and several other piled embankments.

3   MAKING A CHOICE FOR A ROAD CONSTRUCTION METHOD

3.1   Requirements and demands

The choice for a method for road construction on soft and compressible soil has serious implications for both construction phase and operation of the road. Road users want the road to be available as soon as possible and do not accept delays due to maintenance works.

In addition, when building in densely populated areas, owners of nearby underground infrastructure or facilities do not want the construction works to damage their property. However, the principal often has a limited budget that is fixed in an early stage of the project.

These demands often are contradictory, as indicated in Table 1. The table compares the impacts of three categories of construction methods commonly used in the Netherlands. Prefabricated vertical drains are the default option, combined with some extra meters of sand loading as temporary surcharge. The temporary surcharge will reduce creep settlements during the service life of the road. The Beaudrain and IFCO methods use a combination of underpressure in the drains and forced dewatering to apply a temporary surcharge to the soft soil, adding to the effect of the sand surcharge. The principle of both piled embankments and Expanded Polystyrene (EPS) embankments is to prevent the compaction of the soft soil, thus eliminating settlements. Although EPS is also a good construction method, this will not be considered further in this paper.

No single method fulfils all requirements. Piled embankments meet the demands of road users and local stakeholders best, but the construction is generally more expensive for the principal. In the example of the N210, strict requirements were set for creep settlements and the availability of the existing very vulnerable road during construction. With these requirements, and with

TABLE 1 QUALITATIVE COMPARISON OF THREE CATEGORIES OF CONSTRUCTION METHODS FOR ROAD CONSTRUCTION ON SOFT SOIL

Impact of construction method	Traditional: prefabricated vertical drains with temporary surcharge	Beaudrain or IFCO method with temporary surcharge	Piled embankment / EPS embankment
Construction time	long	medium	short
Maintenance	frequent	frequent - medium	none
Construction costs	low	medium	high
Damage to nearby structures	additional measures required	can be avoided by careful execution	unlikely

inclusion of 20 years of maintenance in the contract, the piled embankment came out as the most reliable solution with lowest overall cost. Another example is the piled embankment of the railroad in Houten. Here, very limited construction time was available, and with a vulnerable foundation next to the railroad, an innovative solution was necessary. The piled embankment again came out as the most reliable solution with lowest overall costs.

However, the definition of 'lowest overall cost' is not straightforward. This is illustrated in the next paragraph for the widening of Dutch national highway A2.

3.2 Case study: widening highway A2

Highway A2 is one of the main North-South arteries of the Netherlands, stretching 200 km from Amsterdam to Maastricht. The section between Amsterdam and Utrecht persistently topped the chart of most congested highways, causing an estimated loss of production of 50 million Euros annually. It was decided to widen the existing 2×2 lane highway to 2×4 lanes, by building a new road embankment adjacent to the existing one. Construction has started in 2006, and will be completed in 2010.

The subsoil consists of 5 to 7 m of very weak and compressible peat and organic clay deposits, with occasional buried sand channels of former rivers. Embankments up to 8 m above ground level are required at intersections with other infrastructure, some located at the weakest spots. Construction settlements using conventional methods may be as large as 3 m; with a temporary surcharge of 4 m the total fill thickness reaches 15 m. The time available for embankment filling and preloading is two years. In this time, congestion is worse then ever because of the reduced width of the lanes due to the construction works.

The project took off with the traditional construction method, prefabricated vertical drains and temporary surcharge, selected for most locations. For the high embankments, the Beaudrain and IFCO methods were employed.

However, application of piled embankments could have reduced both construction time and congestion during construction significantly. What would have been the additional costs? And why were the piled embankments not preferred over the traditional methods, given the huge benefits for the road users?

The usual approach to answer these questions is to make a design for all construction options as basis for a cost estimate. This is often avoided because this is time-consuming. However, the last years have brought decision support systems to quantify the impacts of choices, making the selection process more transparent. Computational power has increased such that all calculations only take a few hours.

Most calculations for this case study were done with MRoad, a decision support system dedicated to the selection of construction methods for highways on soft soil (Venmans et al., 2005). Given the standard end user requirements, construction time, subsoil and geometry data, the program automatically performs settlement analysis for eight common construction methods. MRoad predicts maintenance actions for every construction method, and the whole life costs for construction and maintenance. The economic loss due to congestion during construction and maintenance is not automatically calculated, but can be entered manually.

The case study concerns the widening in a 7 km subsection of the project between the Holendrecht River and the town of Vinkeveen. The subsection was characterized by 14 different combinations of subsoil stratigraphy and road geometry. For every combination, the three construction options presented in Table 1 have been compared. The center-to-center distance of the prefabricated vertical drains is 1 m; the thickness of the temporary surcharge is 2 m. The case study assumes the application of timber piles for the piled embankment. The high ground water table in the area is ideal for timber piles, and timber piles may be more cost effective than concrete piles.

Three sets of calculations have been compared, assuming a required construction time of successively 2 years, 1 year and half a year. For every set, a combination of the three construction methods was chosen, assuming zero maintenance in the first 10 years after opening of the road. Fig. 4 shows the results of the calculations.

When the principal requires a construction time of 2 years, most parts of the project can be realized using the traditional construction method and Beaudrain or IFCO methods. Large quantities of sand are required to compensate settlements and for the temporary surcharge, and substantial quantities of temporary surcharge need to be removed again. Piled embankments are necessary only in places where the new embankment crosses the existing road and no settlements are allowed.

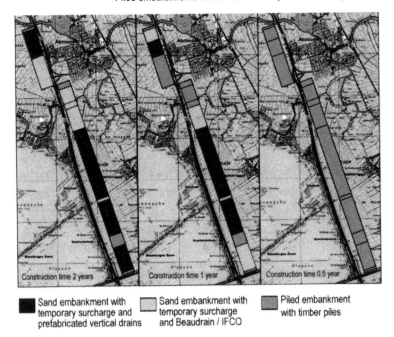

Fig. 4 *Construction methods selected for different construction times*

If the principal would allow only 1 year for construction, the traditional and Beaudrain/IFCO methods cannot fulfil the zero maintenance requirement. Piled embankments are the only solution for transitions to bridges with pile foundations and in areas with strong subsoil heterogeneity.

If only half a year is available for construction, piled embankments are the only option. The logistics of sand transport alone prohibit other construction methods.

The cost comparison learns that the acceleration of construction by one and a half year would have required 8 million Euros. Asking each road user for a fee of 0.25 Euro for one year would have raised this amount. The Dutch economy looses 50 million Euros every year due to congestion on highway A2 between Amsterdam and Utrecht. The subsection in the case study covers one third of this, but it is clear that the profit for the Dutch economy outweighs the additional construction costs for piled embankments. In addition, to be effective as a criterion for selecting construction methods, the 'overall costs' should clearly include the profit for society.

At the same time this – academic – case study was performed, the road administration asked the contractor to reduce the construction time by one year. The new design involved significant lengths of piled embankments, probably resembling the 1 year option in Fig. 4. It was for contractual, not technical reasons that piled embankments were no part of the solution finally adopted. The matter even led to questions in the Dutch parliament. When asked why piled embankments were withheld, the Minister of Transport regarded piled embankments 'too risky'. At the time, highways on piled embankments were performing satisfactory at six locations in the Netherlands.

In reality, the additional costs for acceleration of construction by one year were 33 million Euros. The difference with the case study is largely due to the costs of changing an existing contract. But how should the principal have stimulated short construction times from the onset of the project?

3.3 Stimulating short construction times

Faster construction of the road embankment does not necessarily lead to faster completion of the entire project. The availability of grounds, relocating underground infrastructure, legal procedures and coordination with other construction works may seriously jeopardize the theoretically feasible gain. The present Design & Construct contracts do not give the contractor much room to operate, since the principal has arranged most of these matters before contracting. Accelerated completion of projects will require more focus either on acceleration for the principal, or more freedom for the contractor in dealing with other stakeholders. The last is a shared concern of contractor and principal, and it is only logical that also revenues and the losses are shared.

Various projects in the Netherlands show what can be reached with the changing attitude of the authorities. In the case of N210, the strict requirements for availability and integrity of the existing road, and the responsibility for the maintenance for the first 20 years, forced contractors to offer innovative solutions. Lightweight EPS foam allowed the nearby 2 km long N475 to be reconstructed within a month, a bonus/penalty system stimulating fast construction. The lightweight solution was hardly more expensive than the traditional construction. A piled embankment turned out to be more expensive because the logistics were less suitable for that location. In both cases, the principal gave priority to the interests of the road user. And in both cases, the additional costs of innovative construction methods turned out to be limited.

4   MAKING A DESIGN FOR A EMBANKMENT

At the time that the Dutch Minister of Transport regarded piled embankments 'too risky', several design methods for piled embankments existed. These models tended to give completely different designs, up to a factor 10 difference in design strength of the geosynthetic reinforcement. For the Dutch it was not clear yet what method should be followed.

Since then, the Dutch Task Group 'Design Guideline for Piled Embankments', has introduced the Dutch Design Guideline (Van Eekelen et al., 2010a). Why should this Guideline ensure the Minister that it gives reliable piled embankments?

The Task group evaluated the existing methods mathematically (Van Eekelen and Bezuijen, 2008a) and predictions were compared to finite element calculations (Van Eekelen and Jansen, 2008b and appendix of CUR, 2010) and several field tests. For this purpose, monitoring programs were carried out in the Houten railway (Van Duijnen and Van Eekelen, 2010) and in a pilot piled embankment 'the Kyoto Road' (Van Eekelen et al., 2010b), and the results of the monitoring carried out by the contractor of the N210 were also involved (Haring et al., 2008).

The agreement between finite element calculations and the German EBGEO was very good, the agreement between EBGEO and the field monitoring results was better than with one of the other design methods. Therefore, it was concluded that the arching model given in the EBGEO, was the best available model.

Thus, major parts of the EBGEO had been adopted in the Dutch Guideline. However, constraints are adapted for Dutch circumstances as described in Van Eekelen et al. (2010a). One of them is the reduction of the minimum embankment height, which makes the guideline more suitable for the flat Dutch country. In addition, the guideline has been extended with several chapters, such as the traffic load that has to be taken into account and two approaches for the pile design.

5 CONCLUSIONS AND FUTURE DEVELOPMENTS

While making a decision for a road construction method in the Netherlands, the authorities increasingly often consider more aspects than construction costs only. This leads to application of innovative techniques more frequently. For example, when construction time and reduction of traffic hindrance are paramount, more often innovative constructions, like a piled embankments, come out as the best option. In specific cases a piled embankment is anyway one of the best feasible solutions, for instance if a nearby sensitive construction has to be protected, or the available construction time is limited.

A Dutch guideline has been introduced in the beginning of 2010, which makes it more feasible to apply a piled embankment in the Netherlands. Experiences with the guideline are now being acquired in several piled embankments projects. Meanwhile, measurements continue in the piled embankments of the N210, Houten and (only since a few months) Woerden and laboratory experiments are being carried out in the laboratory of Deltares. In the coming period, these measurements in field and laboratory will be further evaluated and possibly, the Dutch Guideline can be further improved in a few years.

REFERENCES

CUR (2010). Design Guideline for Piled Embankments, CUR report.

Duijnen, P., van and Eekelen, S., J. M. van (2010). Monitoring of a Railway Piled Embankment, *paper nr 186 in the proc. of 9ICG.*

Eekelen, S. J. M. van and Bezuijen, A (2008a). Design of piled embankments, considering the basic starting points of the British Design Guideline, *paper number 315 in the proceedings of EuroGeo4, September 2008, Edinburgh, UK.*

Eekelen, S. J. M. van and Jansen, H (2008b). verslag van een case studie, op weg naar een Nederlandse ontwerprichtlijn voor paalmatrassen, *Geotechniek* July 2008, pp. 66-71.

Eekelen, S. J. M. van, Jansen, H. L., Duijnen, P. G. van, De Kant, M., Dalen, J. H. van, Brugman, M. H. A., Stoel, A. E. C. van der, Peters, M. G. J. M (2010a). The Dutch Design Guideline for Piled Embankments, *paper nr 120 in the proc. of 9ICG*, Brazil, 2010.

Eekelen, S. J. M. van, Bezuijen, A. and Alexiew, D. (2010b). The Kyoto Road, monitoring a piled embankment, comparing $3^1/_2$ years of measurements with design calculations, *paper nr 461 in the proc. of 9ICG*, Brazil, 2010.

EBGEO: Bewehrte Erdkörper auf punkt- und linienformigen Traggliedern, Empfehlung für den Entwurf und die Berechnung von Erdkörper mit Bewehrungen als Geokunststoffen, *Ausgabe 02/2009, Kapitel 6.9. Deutsche Gesellschaft für Geotechnik e.V. Arbeitskreis 5.2.*

Eurocode NEN-EN (1990). Grondslagen van het constructief ontwerp.

Haring, W., Profittlich, M. and Hangen, H. (2008). Reconstruction of the national road N210 Bergambacht to Krimpen a.d. IJssel, nl: design approach, construction experiences and measurement results, *4th European Geosynthetics Conf., September 2008, Edinburgh, UK.*

Heemstra, J. (2008). Wat wij nu nog van Keverling Buisman kunnen leren: De betekenis van klassieke matrassen in de wegenbouw voor de paalmatras van vandaag. *GeoKunst* juli 2008, nr. 2, pp 54-57.

Keverling Buisman, A. S. (1937). Letter to the Dutch Province of South-Holland, February 8th, 1937, available at Deltares.

Venmans, A. A. M., Förster, U., Hooimeijer, R. H. (2005). Integral design of motorways on soft soil on the basis of whole life costs. *Proc. of the 16th Int. Conf. on Soil Mechanics and Geotechnical Engineering.* Osaka, Japan, Vol. 4:2867-2870. Rotterdam: Millpress.

Consolidation settlement of stone column-reinforced foundations in soft soils

Jie Han, Ph.D., PE
Department of Civil, Environmental, and Architectural Engineering, the University of Kansas, Lawrence, Kansas, USA

KEYWORDS consolidation, settlement, soft soil, stone column

ABSTRACT Stone columns have been successfully used to reduce settlement and accelerate consolidation of soft soils. Different methods have been proposed to calculate the settlement and rate of consolidation of stone column-reinforced foundations. A field embankment over stone column-reinforced soft clay was selected from the literature to examine some of these methods. Discussion is presented in this paper on the reasonableness and accuracy of these methods. Recommendations are made for the calculation of consolidation settlement of stone column-reinforced foundations in soft soils.

1 INTRODUCTION

Stone columns, as one of the ground improvement technologies, have been commonly and successfully used to reduce settlement and accelerate consolidation of soft soils. The reduction of the settlement is mainly attributed to the higher modulus of the columns (Aboshi et al., 1979; Priebe, 1995; Pulko and Majes, 2005). The acceleration of the consolidation of soft soils is attributed to both drainage and higher modulus of the columns (Han and Ye, 2001, 2002; Tan et al., 2008; Castro and Sagaseta, 2009; Xie et al., 2009a and 2009b; and Wang, 2009). There have been a number of projects in the field demonstrating the reduced settlement and accelerated consolidation of stone column-reinforced soft soils, for example, Munfakah et al. (1983), Han and Ye (1992), and Logar et al. (2005). However, some field studies, for example, Garga and Medeiros (1995), showed that stone columns might not work as drains to accelerate the consolidation of soft soils. Barksdale (1987), Han and Ye (2002), and Xie et al. (2009a) attributed the reduced benefits of stone columns in consolidation to the contamination of stones with fine-grained soil, the disturbance of the surrounding soil, and the intrusion of stones into the surrounding soil. Boulanger et al. (1998) reviewed three case studies of stone column projects and found that the intermixing of imported aggregate with native soil could reduce the permeability of stone columns to less or much less than 100 times the permeability of the native soil.

Several design methods have been proposed in the past by different researchers to calculate the settlement and rate of consolidation for stone column-reinforced foundations. However, these methods have not been well evaluated against field performance of stone column-reinforced foundations. A case study from the literature was selected for this purpose.

2   REVIEW OF DESIGN METHODS

2.1 Methods for settlement calculation

2.1.1 Stress reduction factor

A stress reduction factor was proposed by Aboshi et al. (1979) to calculate the stress on the surrounding soil to sand columns as follows:

$$\sigma_s = \mu_s p \qquad (1)$$

$$\mu_s = \frac{1}{1 + a_s (n - 1)} \qquad (2)$$

where

σ_s - the stress on the surrounding soil (see Fig. 1),

p - the average stress applied on the stone column-reinforced foundation,

μ_s - the stress reduction factor, defined as the ratio of the stress on the surrounding soil to the average applied stress,

a_s - the area replacement ratio, defined as the cross-section area of columns divided by the total treated area, and

n - the stress concentration ratio, defined as the ratio of the stress on the column to that on the soil (see Fig. 1).

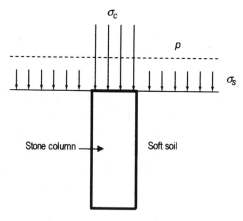

FIG. 1 *Stress distribution on a stone column-reinforced foundation*

Even though this formula was first proposed for sand columns, it is applicable to stone columns. The reduced stress on the surrounding soil can be used to calculate the settlement of the stone column-reinforced foundation using the Boussinesq solution or other simplified methods, which is the same as that for the unreinforced foundation. The settlement of the stone column-reinforced foundation can be expressed as follows:

$$s_{sc} = \mu_s s_s \qquad (3)$$

where

s_{sc} - the consolidation settlement of the stone column-reinforced foundation,

s_s - the consolidation settlement of the unreinforced foundation.

In this method, one of the key tasks is to estimate the stress concentration ratio. Field tests have shown that the stress concentration ratio mostly ranged from 2.0 to 5.0. From laboratory tests, Barksdale and Bachus (1983) established an empirical relationship between the stress concentration ratio and the modulus ratio of the stone columns to the soft soil as shown in Fig. 2. The average stress concentration ratio, n, in Fig. 2 can be expressed in a formula as follows:

$$n = 1 + 0.217\left(\frac{E_c}{E_s} - 1\right) \qquad (4)$$

where

E_c - the elastic modulus of the columns,

E_s - the elastic modulus of the soft soil.

Based on the measured stress concentration ratios (mostly 2.0 to 5.0) in the field, the modulus ratio should be limited to 20 for the above equation in the design. Huang and Han (2008) found that deep mixed columns could not mobilize their strengths up to 20 times the soft soil strengths under embankments because of plastic deformation of the embankment fill.

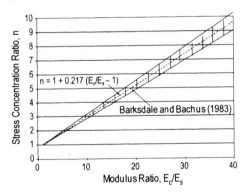

FIG. 2 *Stress concentration ratio versus modulus ratio*

2.1.2 Improvement factor

Priebe (1995) proposed a basic improvement factor for stone column-reinforced foundations by a vibro-replacement method considering rigid and incompressible columns with a bulging over the column length. When the

Poisson's ratio of the soil equals to 1/3, a simplified solution for the basic improvement factor, I_f, can be expressed as follows:

$$I_f = 1 + a_s \left[\frac{5 - a_s}{4(1 - a_s) \tan^2 \left(45^\circ - \phi_c/2\right)} - 1 \right] \tag{5}$$

where

ϕ_c - the friction angle of the column. The design friction angle of stone columns mostly ranges from 35° to 45°.

The improvement factor can be used to calculate the settlement of the stone column-reinforced foundation based on the settlement of the unreinforced foundation as follows:

$$S_{sc} = \frac{S_s}{I_f} \tag{6}$$

Priebe (1995) also suggested the consideration of the column compressibility and overburden pressure in addition to the basic improvement factor. The formulae and design charts for such consideration can be found in Priebe (1995).

2.1.3 Elastic-plastic solution

Pulko and Majes (2005) and Castro and Sagaseta (2009) proposed their methods for calculating the settlement of the stone column-reinforced foundation based on elastic-plastic constitutive models. In their methods, the soft soil is assumed to be linearly elastic while the stone columns are assumed to be linearly elastic-perfectly plastic following the Mohr-Coulomb failure criterion with a constant dilatancy angle. The plasticity starts at the upper portion of the column and can extend deeper to the whole length of the column with an increase of the applied load. Details on these solutions can be found in Pulko and Majes (2005) and Castro and Sagaseta (2009).

2.2 Methods for consolidation rate calculation

2.2.1 Solutions based on one-dimensional deformation

The author (Han, 1989) first developed the simplified solutions for the consolidation rates of stone column-reinforced foundations with and without well resistance and smear effects during his Master's study in China from 1986 to 1989. Later he published these solutions with his advisor on two international journals with some improvement (Han and Ye, 2001; Han and Ye, 2002). These solutions were developed based on the assumptions of equal strain and one-dimensional deformation of the column and the soil. Modified coefficients of consolidation of soft soil in vertical and radial directions were introduced and are presented below:

$$c_{vm} = c_v \left(1 + n \frac{1}{N^2 - 1}\right) = c_v \left(1 + n \frac{a_s}{1 - a_s}\right)$$

$$c_{rm} = c_r \left(1 + n \frac{1}{N^2 - 1}\right) = c_r \left(1 + n \frac{a_s}{1 - a_s}\right) \tag{7}$$

where

c_v and c_r - the coefficients of consolidation of soft soil in vertical and radial directions, respectively,

c_{vm} and c_{rm} - the modified coefficients of consolidation of soft soil in vertical and radial directions, respectively, and

N - the diameter ratio, defined as the ratio of the influence diameter to the column diameter in a unit cell.

The modified coefficients of consolidation of soft soil account for the contribution of the stress concentration on the columns due to the modulus difference between column and soil. The modified coefficient of consolidation of soft soil in the vertical direction can be used to calculate the time factor for Terzaghi's solution for the rate of one-dimensional consolidation:

$$T_{vm} = \frac{c_{vm} t}{H^2} \tag{8}$$

where

t - the time,

H - the longest drainage distance in the vertical direction.

The modified coefficients of consolidation of soft soil in the radial direction can be used to calculate the time factor for Hansbo's solution (Hansbo, 1981) for the rate of radial consolidation:

$$T_{rm} = \frac{c_{rm} t}{d_e^2} \tag{9}$$

where

d_e - the effective or influence diameter of an individual column in a unit cell.

Han and Ye (2002) developed the following equation to calculate the rate of consolidation of the stone column-reinforced foundation considering smear and well resistance effects:

$$U = 1 - e^{-\frac{8}{F'_m} T_{rm}} \tag{10}$$

$$F'_m = \frac{N^2}{N^2 - 1}\left(\ln \frac{N}{S} + \frac{k_r}{k_s}\ln S - \frac{3}{4}\right) + \frac{S^2}{N^2 - 1}\left(1 - \frac{k_r}{k_s}\right)\left(1 - \frac{S^2}{4N^2}\right)$$

$$+ \frac{k_r}{k_s}\frac{1}{N^2 - 1}\left(1 - \frac{1}{4N^2}\right) + \frac{32}{\pi^2}\left(\frac{k_r}{k_c}\right)\left(\frac{H}{d_c}\right)^2 \tag{11}$$

where

$S = d_s/d_c$ - the diameter ratio of the smeared zone to the stone column,

d_s - the diameter of the smeared zone,

d_c - the diameter of the column,

k_r - the radial permeability of the undisturbed surrounding soil, and

k_s - the radial permeability of the smeared soil.

Field observations showed that stones were pushed into the surrounding soft soil and intermixed with the soft soil during the stone column installation. Baez and Martin (1995) and Boulanger et al. (1998) indicated that the intrusion of native soil into stone columns could reach 20% by weight. Field injection tests indicated the ratio of the permeability of stone columns to that of the native soil ranging from 15 to 40 while laboratory tests indicated this ratio from 40 to 100. The difference might result from the smear effect, which was included in the field tests, but not in the laboratory tests.

Xie et al. (2009b) derived a solution for the consolidation rate of the stone column-reinforced foundation considering the variation of the permeability of the surrounding soil from the interface between the column and the surrounding soil to the boundary of a unit cell. Determination of such variations in the field is a challenging task. In addition, Xie et al. (2009a) developed another solution considering the variations of permeability and total stresses with depth and time. The variation of the total stress with depth is more important for a small-area loading. Xie et al. (2009a) also considered a step or ramp loading situation, which is useful to simulate the construction load, such as the filling of an embankment.

In addition to a step or ramp loading, Wang (2009) developed a solution for the rate of consolidation for the stone column-reinforced foundation subjected to a cyclic loading, which may be useful to simulate traffic loading.

Both Xie et al. (2009a) and Wang (2009) compared their solutions to that of Han and Ye (2002) under an instantaneous load and resulted in a good agreement. All the above-mentioned solutions were developed based on the assumptions of equal strain and one-dimensional vertical elastic deformation in the column and the soil.

2.2.2 Solutions considering lateral and/or plastic deformation

Castro and Sagaseta (2009) developed their solutions considering lateral deformation of stone columns under a vertical load. The lateral deformation of the column reduces the load carried by the column and slows down the rate of consolidation. In addition, Castro and Sagaseta (2009) treated the stone column as a linearly elastic perfectly plastic material. The upper portion of the column can yield under a certain load and the plastic zone can extend deeper to the whole length of the column when the applied load is increased. They also developed modified coefficients of consolidation based on elastic and plastic deformations. However, the Castro and Sagaseta (2009) solutions did not consider smear and well resistance effects on the rate of consolidation of the stone column-reinforced foundation.

3   EVALUATION AND DISCUSSION

The field study reported in Tan et al. (2008) was selected in this study to evaluate the methods discussed above. A 40 m wide and 1.8 m high embankment was constructed over a 5 m thick soft clay underlain by a stiff clay. Above the soft soil, there was a 1 m thick fill, which was provided as a construction platform and drainage layer. The groundwater table was 1 m below the ground surface. The material properties used by Tan et al. (2008) are provided in Table 1. The stone columns installed in this project had a diameter of 0.8 m and a length of 6 m with a square pattern at spacing of 2.4 m. The area replacement ratio is 0.087.

TABLE 1   MATERIAL PROPERTIES (ADAPTED FROM TAN ET AL., 2008)

Material	γ (kN/m^3)	E (MPa)	k_r (m/s)	k_v (m/s)	c' (kPa)	ϕ' ($^\circ$)
Embankment fill	18	15	1.16×10^{-5}	1.16×10^{-5}	3	23
Crust	17	15	3.47×10^{-7}	1.16×10^{-7}	3	28
Soft clay	15	1.1	3.47×10^{-9}	1.16×10^{-9}	1	20
Stiff clay	20	40	3.47×10^{-9}	1.16×10^{-9}	3	30
Stone column	19	30			5	40

γ = unit weight (unsaturated above or saturated below groundwater table), E = elastic modulus, k_r = radial permeability, k_v = vertical permeability, c' = effective cohesion, and ϕ' = effective friction angle.

The permeability of stone columns was arbitrarily chosen by Tan et al. (2008) to be 10,000 times that of the soft clay. Based on the earlier discussion, the permeability of stone columns could be 15 to 100 times that of the native soil, which is much lower than what Tan et al. (2008) assumed. The permeability of a granular drain with fine contents can be estimated using the formula in the FHWA *Highway Subdrainage Design* manual (Moulton, 1980):

$$k = \frac{2.19 \, (D_{10})^{1.478} \, n^{6.654}}{(P_{200})^{0.597}} \tag{12}$$

where
   k - the permeability of the granular drain (i.e., stone column in this study) (unit: m/s),
   D_{10} - the effective grain size corresponding to 10% passing the sieve size (unit: mm),
   P_{200} - the percentage of particles passing U.S. No. 200 sieve,
   n - porosity = $1 - \frac{\gamma_d}{\gamma_w G}$,

γ_d - the dry unit weight of the stone column,

γ_w - the unit weight of water, and

G - the specific gravity (assumed = 2.70).

As discussed earlier, 20% native soil (soft clay) could be mixed into the stone columns, i.e., $P_{200} = 20$. Considering the size of clay particles and the minimum effective grain size used the FHWA Nomogram of 5μm, $D_{10} = 0.005$ mm. The dry unit weight of the stone columns of 15.7 kN/m^3 was used in the calculation. The calculated permeability of the stone columns is 3.64×10^{-7} m/s, which is approximately 314 times the permeability of the soft clay in the vertical direction. This ratio is higher than that obtained by Baez and Martin (1995) and Boulanger et al. (1998) from field studies, in which stone columns were installed in sandy or silty soils. The permeability of stone columns measured in the field resulted from the combined effect of well resistance and smear. However, it is a challenging task to determine the zone of smear and the permeability of the smear zone. For convenience, the effect of smear is not considered herein, but considered as part of well resistance with the reduced permeability of the stone column.

From Table 1, the elastic modulus ratio of the column to the soil is 36.4. As discussed earlier, a maximum modulus ratio of 20 can be used for Fig. 2 or Eq. 4 to determine the stress concentration ratio (i.e., $n = 5.1$).

To calculate the settlement at the centerline of the embankment, a constrained modulus of the soft clay, D, should be used and can be calculated as follows:

$$D = \frac{(1 - v) E}{(1 + v)(1 - 2v)} \tag{13}$$

The calculated constrained modulus of the stone columns is 1.5 MPa.

The total consolidation settlement of the embankment can be calculated based on the one-dimensional compression theory, i.e.:

$$s_s = \sum \frac{\Delta\sigma_{zi} h_i}{D} \tag{14}$$

where

s_s - the consolidation settlement at the base of the embankment,

$\Delta\sigma_{zi}$ - the additional vertical stress at the depth of z_i due to the embankment loading,

h_i - the thickness of the ith sub-layer.

The calculated consolidation settlement for the unreinforced foundation is 109.4 mm. The stress reduction factor in Eq. 4 can be calculated as 0.735. The total consolidation settlement of the stone column-reinforced foundation using the stress reduction method is $0.735 \times 109.4 = 80.5$ mm.

Eq. 5 can be used to calculate the basic improvement factor as 1.45 and therefore the total consolidation settlement of the stone column-reinforced foundation using the improvement factor method is $109.4/1.45 = 75.4$ mm. However, if the column compressibility and the overburden are considered, the final improvement factor is increased to 2.19, which results in a settlement of 50 mm. Since the column has a large elastic modulus, its effect on the final improvement factor is minimal. However, the inclusion of the overburden stress has a significant effect.

The Castro and Sagaseta (2009) method calculated the settlement of the stone column-reinforced foundation at 76 mm. Their method also indicated the whole column in a plastic state.

Except the calculated settlement (50 mm), based on the final improvement factor, these three methods result in close settlements, which are also close to the measurement (approximately 80 mm) in the field. This comparison shows that the basic improvement factor method calculated more reasonable settlement than the final improvement factor method as compared with the field data. The average total consolidation settlement from these three methods (excluding 50 mm from the total improvement factor method) is 77.3 mm, which will be used in the later calculation.

Based on the provided material properties, the coefficients of consolidation of the soft soil in the vertical and radial directions were calculated as 1.75×10^{-7} and 5.24×10^{-7} m^2/s, respectively. The modified coefficients of consolidation of the stone column-reinforced foundation in the vertical and radial directions were calculated as 2.61×10^{-7} and 7.80×10^{-7} m^2/s, respectively, using Eq. 7.

Eqs. 8 to 11 can be used to calculate the rates of consolidation in both vertical and radial directions in terms of time. Considering a combined effect of radial and vertical flows, the overall rate of consolidation can be calculated using the following formula (Carillo, 1942):

$$U_{rv} = 1 - (1 - U_r)(1 - U_v) \qquad (15)$$

where
U_{rv} - the overall rate of consolidation in both directions,
U_r - the rate of consolidation in the radial direction, and
U_v - the rate of consolidation in the vertical direction.
The consolidation settlement at time, t, can be calculated as follows:

$$s_{sc}(t) = U_{rv} \cdot s_{sc} \qquad (16)$$

Fig. 3 shows the measured and calculated settlement-time curves at the centerline of the embankment. It is shown that the solution without considering well resistance overestimated the rate of consolidation while the solution considering well resistance resulted in a settlement-time curve in good agreement with the measured. This comparison demonstrates the importance

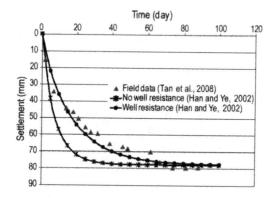

FIG. 3 *Measured and calculated settlement-time curves*

of considering well resistance in calculating the rate of consolidation for the stone column-reinforced foundation.

4   RECOMMENDATIONS

The following procedure is recommended for calculating the consolidation settlement of stone column-reinforced foundations at time, t:

1. Fig. 2 or Eq. 4 can be used to determine a stress concentration ratio based on the modulus ratio of stone columns and soil. The modulus ratio should be limited to 20 during this calculation. The stress concentration ratio mostly ranges from 2.0 to 5.0.

2. Eq. 12 can be used to estimate the permeability of stone columns considering the intermixing of native soil with aggregates. Tipically, 20% native soil can be mixed into stone columns.

3. A constrained modulus from odometer tests or estimated using Eq. 13 should be used for the calculation of the consolidation settlement at the center of the foundation.

4. The settlement of the unreinforced foundation can be calculated using conventional methods including Eq. 14.

5. A stress reduction ratio or improvement factor method or the Castro and Sagaseta (2009) method can be used to calculate the settlement of a stone column-reinforced foundation.

6. The rate of consolidation of the stone column-reinforced foundation can be estimated using Han and Ye (2002) or other methods considering well resistance.

7. The settlement at a time can be calculated using Eq. 16.

5 CONCLUSIONS

This paper reviews the methods available to calculate the total consolidation settlement and rate of consolidation of stone column-reinforced foundations. The stress reduction factor method, the basic improvement factor method, and the Castro and Sagaseta (2009) method calculated similar total consolidation settlements as the field measurement. The consideration of the overburden stress effect in the calculation of the final improvement factor underestimated the settlement. Field studies showed that the permeability of stone columns ranged from 15 to 100 times that of the native sandy or silty soil. The analysis in this study confirmed that the stone columns in the selected case study had lower permeability, which is approximately 314 times that of the native clay. Therefore, it is important to consider well resistance in the calculation of the rate of consolidation for a stone column-reinforced foundation. The comparison shows that the solution developed by Han and Ye (2002) considering well resistance can reasonably predict the settlement-time curve for the stone column-reinforced foundation. A design procedure is recommended in this paper to help geotechnical engineers calculate the consolidation settlement of stone column-reinforced foundations in soft soils.

ACKNOWLEDGEMENT

Dr. Jorge Castro at Universidad de Cantabria in Spain provided a spreadsheet of the Castro and Sagaseta (2009) method, which was used in this study. His help is greatly appreciated.

REFERENCES

Aboshi, H., Ichimoto, E., Enoki, M., and Harada, K. (1979). The compozer – a method to improve characteristics of soft clays by inclusion of large diameter sand columns. *Proceedings of the International Conference on Soil Reinforcement*, E.N.P.C., 1, Paris, France, pp. 211-216.

Barksdale, R. D. and Bachus, R. C. (1983). *Design and Construction of Stone Columns*, FHWA/RD-83/026, 194p.

Barksdale, R. D. (1987). Applications of the state of the art of stone columns-liquefaction, local bearing failure, and example calculations, *Technical Report REMR-GT-7*, The Georgia Institute of Technology, 90p.

Boulanger, R. W., Idriss, I. M., Stewart, D. P., Hashash, Y., and Schmidt, B. (1998). Drainage capacity of stone columns or gravel drains for mitigating liquefaction. *Proceedings of a Geotechnical Earthquake Engineering and Soil Dynamics III*, ASCE, pp. 678-690.

Carillo, N. (1942). Simple two and three dimensional cases in the theory of consolidation of soils. J. Math. Phys., Vol. 21(1), pp. 1-5.

Castro, J. and Sagaseta, C. (2009). Consolidation around stone columns – influence of column deformation. *International Journal for Numerical and Analytical Methods in Geomechanics*, Vol. 33(7), pp. 851-877.

Garga, V. K. and Medeiros, L. V. (1995). Field performance of the port of Sepetiba test fills. *Canadian Geotechnical Journal*, Vol. 32, pp. 106-121.

Han, J. and Ye, S. L. (1992). Settlement analysis of buildings on the soft clays stabilized by stone columns, *Proceedings of the International Conference on Soil Improvement and Pile Foundations*, Nanjing, China, pp. 446–451.

Han, J. (1989). "Experimental and theoretical studies of soft soils reinforced by stone columns." *MS thesis (in Chinese)*, Tongji University, P.R. China, 69p.

Han, J. and Ye, S. L. (2001). Simplified method for consolidation rate of stone column reinforced foundations. *Journal of Geotechnical and Geoenvironmental Engineering*, ASCE, Vol. 127(7), pp. 597-603.

Han, J. and Ye, S. L. (2002). A theoretical solution for consolidation rates of stone column-reinforced foundations accounting for smear and well resistance effects. *The International Journal of Geomechanics*, Vol. 2(2), pp. 135-151.

Hansbo, S. (1981). Consolidation of fine-grained soils by prefabricated drains, *Proc. Of 10th Int. Conf. On Soil Mech. and Found. Engg.*, Stockholm, Vol. 3, pp. 677-682.

Huang, J. and Han, J. (2008). "Critical height of deep mixed column-supported embankment under an undrained condition." *GeoCongress 2008: Geosustainability and Geohazard Mitigation*, ASCE, pp. 638-645.

Logar, J., Majes, B., Turk, M. R., and Locniskar, A. (2005). *Proceedings of the 16th International Conference on Soil Mechanics and Geotechnical Engineering*, Osaka, Japan, pp. 1889-1892.

Moulton, L. K. (1980). *Highway Subdrainage Design*, Report FHWA-TS-80-224, Offices of Research and Development, Federal Highway Administration, Washington DC.

Munfakah, G. A., Sarkar, S. K., and R. J. Castelli, R. J. (1983). Performance of a test embankment founded on stone columns, *Proceedings of the International Conference on Advances in Pilings and Ground Treatment for Foundations*, London, pp. 259–265.

Priebe, H. J. (1995). The design of vibro replacement. *Ground Engineering*, December, 31-37.

Pulko, B. and Majes, B. (2005). Simple and accurate prediction of settlements of stone column reinforced soil. *Proceedings of the 16th International Conference on Soil Mechanics and Geotechnical Engineering*, Osaka, Japan, pp. 1401-1404.

Tan, S. A., Tjahyono, S., and Oo, K. K. (2008). Simplified plane-strain modeling of stone-column reinforced ground. *Journal of Geotechnical and Geoenvironmental Engineering*, ASCE, Vol. 134(2), pp. 185-194.

Wang, G. (2009). Consolidation of soft clay foundations reinforced by stone columns under time-dependent loadings. *Journal of Geotechnical and Geoenvironmental Engineering*, ASCE, Vol. 135(12), pp. 1922-1931.

Xie, K.-H., Lu, M.-M., Hu, A.-F., and Chen, G.-H. (2009a). A general theoretical solution for the consolidation of a composite foundation. *Computers and Geotechnics*, Vol. 36, pp. 24-30.

Xie, K.-H., Lu, M.-M., and Liu, G.-B. (2009b). Equal strain consolidation for stone columns reinforced foundation. *International Journal for Numerical and Analytical Methods in Geomechanics*, Vol. 33(15), pp. 1721-1735.

Modeling of ground improvement using stone columns

Weber, T. M.
Studer Engineering, Zurich, Switzerland

KEYWORDS ground improvement, stone columns, soil structure interaction, centrifuge modeling, numerical modeling

ABSTRACT Stone columns are used in practice for ground improvement of soft subsoil in order to accelerate consolidation, reduce compressibility, and increase strength. The presented paper investigates the behavior of stone columns constructed in soft clay and deepens the understanding of the system behavior of the improved ground and effects taking place during column construction. In order to gain a better understanding of the interaction within the improved ground, physical investigations and numerical analysis are being conducted by means of centrifuge and finite element modeling. Particular emphasis is given to the modeling and simulation of column installation. Centrifuge tests are performed using a stone column installation tool for in-flight column construction in order to reproduce prototype situation. An axis symmetric finite element model is developed to simulate the construction of a stone column in a unit cell approach. Further, a plane strain finite element model is used to back analyze the behavior of an embankment on improved ground. Measurements from centrifuge tests are taken for comparison and calibration of numerical models. The analyses of test data show the change in stress state and clay structure around stone column due to their installation and the influence on clay behavior. Further, the numerical model shows a load dependency and the development of stress concentration of the improved ground. The consolidation characteristics of clay also changes due to column installation, which has an influence on the time settlement behavior on the improved ground.

1 INTRODUCTION

Construction of stone columns in soft soil is a common economical ground improvement method when shear strength increase, settlement reduction and acceleration of consolidation are needed. In particular, when less settlement-sensitive structures such as embankments are designed, the more cost effective method of stone columns is often preferred to other methods like pilling. One of the most frequently used installation methods for stone columns is the bottom feed vibro-displacement method (Barksdale and Bachus, 1983).

The settlement prediction of a structure on improved ground is a challenging task, since the behavior is governed by complicated interaction between soil, stone columns and structure. Simplification in analysis is needed.

In recent years the finite element method alternatively gained spread for design of ground improvement measures. In comparison to analytical design, the numerical simulation is greater in effort and time. Also, simplifications need to be made, particularly when using 2D numerical analysis. However, assumptions have to be made for more sophisticated 3D numerical models to reproduce the stress history and realistic system response.

In order to overcome the deficiencies of the existing methods and to gain a better understanding of the interaction behavior of stone columns, physical investigations are being carried out by using the Zurich geotechnical drum centrifuge (Springman et al., 2001). An installation tool for stone columns construction has been developed and several tests have been conducted modeling one-dimensional consolidation of column grids and embankments on improved clay. The results of selected tests were taken as the basis for axis symmetric and 2-dimensional finite element models. The approaches in centrifuge and numerical modeling contribute to a better understanding of the behavior of stone columns in soft ground.

2   CENTRIFUGE MODELING

In order to model the stone column installation process, an installation tool has been developed for the geotechnical drum centrifuge to construct columns in flight. This tool reproduces the construction process of a displacement column using a dry bottom feed method similar to the vibro-displacement method (Weber et al., 2006).

The construction process of a stone column is shown schematically in Fig. 1. The main parts of installation tool are the filling tube with a transition piece mounted to the working table. Also a laser scanning device is fixed on top of the tool in order to measure heave or settlement of the model surface. In order to prevent plugging of the filling tube with clay during installation of the column, a lost tip system is used. A drawing pin is stuck onto the surface of the clay model before column installation, which closes the opening of the filling tube when the tool is driven into the clay. When the desired depth is reached, sand is filled into the inlet sand hose from outside the centrifuge. Compaction of the stone column is achieved during withdrawal through re-insertion of the installation tool into the soil.

Centrifuge model tests were performed at 50-times gravity. Referring to scaling laws, the model was kept the same representing a prototype structure (model scale in brackets) of 7 m (140 mm) clay layer depth with a stone column grid spacing of 1.7×1.7 m (34×34 mm) and an average column length of 5 m (100 mm) with 0.6 m (12 mm) column diameter. This gives an area improvement ratio of 10 %. The grid layout of the stone column and the clay model are shown in Fig. 2. The test was conducted in a modeling container, which was divided into 2 sections. In one section, the soft clay was improved with stone columns, while in the other section the clay was left unimproved for comparison. Natural clay material from Birmensdorf (Switzerland) was used for the clay model. The material is classified as highly plastic clay (CH, clay

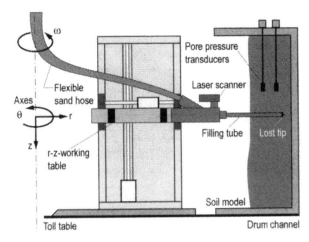

FIG. 1 *Construction of stone columns in the drum centrifuge (Weber et al., 2006)*

FIG. 2 *Preparation of the stone column grid, separation of model container into two parts, left side unimproved and right side improved (Weber et al., 2006)*

particle content 42%, $w_l = 58\%$, $w_p = 19\%$, $I_p = 39\%$, $C_c = 0.44$, $C_s = 0.05$ (Weber, 2008).

During column construction, pore water pressure was measured within the column grid and later used for analysis and back-calculation in a finite element model (see Fig. 7). Fig. 3 shows the excavation of the soil model after the centrifuge test. The embankment is schematically drawn on top of the clay layer because the loose embankment material collapses when the drum centrifuge stops.

After the performance of the centrifuge model test, post test investigations of the clay model were performed. Specimens were analyzed using the methods of Environmental Scanning Electron Microscopy (ESEM) and Mercury Intrusion Porosimetry (MIP).

Fig. 4a shows the ESEM pictures of the structure of the silty clay close to the stone column. The silty soil particles are aligned parallel to each other as

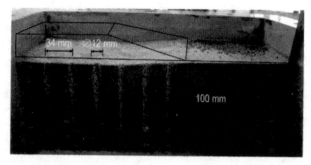

FIG. 3 *Cross section of soil model after the centrifuge test (Weber et al., 2006)*

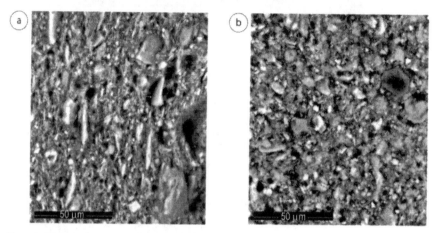

FIG. 4 *ESEM pictures, a) in Zone 2 – silty clay structure adjacent to the edge of the stone column at a radius of approximately 7 mm, b) in Zone 3 – clay structure at 2 mm beyond the edge of the stone column (for zones see Fig. 5) (Weber et al., 2010)*

well as being oriented parallel to the edge of the stone column. The reason for this orientation can be explained by high shearing and remolding due to the installation process. The thickness of this zone 2 is about 2 mm, which corresponds to about 1/3 of the final stone column radius. Fig. 4b shows the ESEM picture of the clay structure just 2 mm outside the edge of the column, where there is no visible reorientation of the clay particles. Moving further away from the column, no observable change in the silty clay structure is apparent from the ESEM pictures.

Reporting the data from the Mercury Intrusion Porosimetry, the results show a trend that the porosity and density change with radius and lead to the conclusion that the clay has been compacted in the vicinity of the stone column. The clay that is affected by the installation process appears to reach to a distance of about 15 mm from the column axis, equivalent to 2.5 times the nominal column radius of 6 mm. Inspection of ESEM pictures reveals no visual difference in clay structure beyond 2 mm from the column, but the clay has experienced densification according to the MIP measurements in zone 3. Beyond the densification zone, there is no observable compaction or change in the clay structure in zone 4.

Fig. 5 shows the summary of the results from smear zone identification around stone columns from centrifuge model tests. Discrete measurements are given at model scale in terms of radius r to the stone column axis and relative measurements are also specified with normalized radius to column radius r/r_w. There is also a zone 1 visible where sand particles penetrate into the clay matrix. This zone is not investigated further.

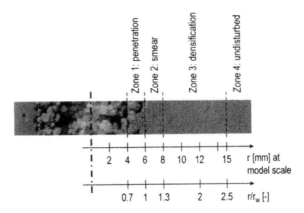

FIG. 5 *Zones of disturbance due to stone column installation (Weber et al., 2010)*

3   NUMERICAL MODELING

The installation of stone columns changes the stress state and the structure in the ground. Representing a 3D stone column grid in 2D is usually represented by a series of parallel trenches. The stiffness as well as the consolidation behavior of both soft soil and coarse grained inclusion needs to be adapted for a plane strain situation in order to model the deformation behavior and drainage conditions for consolidation correctly. Hird et al. (1992) and Indraratna and Redana (1997) recommend how to perform a conversion of permeability in order to match the consolidation behavior. These transformations are also applicable to smear effects, which need to be considered since the drainage conditions and consolidation behavior will be significantly affected by smear when studying time dependent behavior.

The system behavior of a single stone column and a stone column grid was analyzed using 2-dimensional finite element models, one axis symmetric model simulating a unit cell and a plane strain model. Fig. 6 shows both finite element models employed in reference to the centrifuge models test.

The dimensions are given in prototype scale, also the location of the stress points where the development of pore water pressure is registered. As software, PLAXIS V8 (Brinkgreve, 2002) is used for the calculation with 15 node elements. The constitutive models used are Hardening Soil for the clay material and Mohr-Coulomb for the stone columns. In the plane strain model, a smear zone is introduced around the stone columns in order to take the effects of column installation on the drainage situation into account.

A crucial part of the axis symmetric model is the simulation of the stone column installation. The procedure chosen represents a stepwise cavity opening along the symmetry axis. The stone column is divided into 6 segments. First, the inner strip of the top column segment is displaced horizontally to a radius of 0.25 m. As second step, the bottom of the first segment and the top of the second segment is displaced horizontally. The procedure is repeated

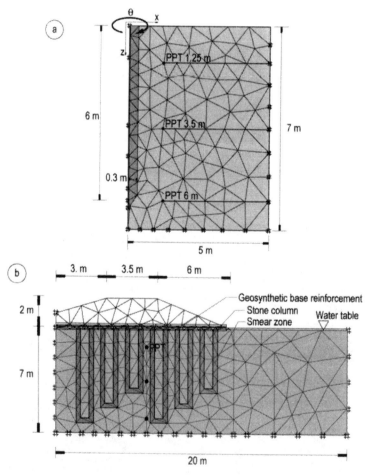

FIG. 6 *2D Finite Element models, a) axis-symmetric unit cell, b) plane strain (Weber, 2008)*

until final depth of the stone column is reached. The compaction during withdrawal of the poker is modeled by further stepwise horizontal displacement of additional 0.05 m to the final column radius of 0.30 m from the bottom segment to the top.

The development of excess pore water pressure is given in Fig. 7 for the measurements from the centrifuge test and the numerical simulation in comparison. The construction process took 400 s in the centrifuge test representing 11.6 days in prototype scale (scaling factor $1/n^2$). The key features of excess pore pressure development during column construction could be modeled; (1) insertion of the poker, (2) full penetration, filling in of sand and start of the compaction sequence until at (3) the installation tool reaches the surface. There is an increase of excess pore water pressure until the tip of the tool passes the level of the stress point during insertion phase and a further increase during compaction phase.

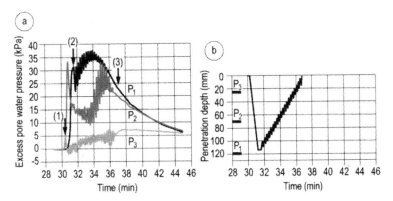

FIG. 7 *Construction of a single stone column, a) excess pore water pressure development in the clay model during column construction in the centrifuge and numerical model, pore pressure transducers P1, P2 and P3 at distinct depths, b) penetration path of the installation tool into the clay model during column construction (Weber et al., 2008)*

Fig. 8 shows the result of stress analysis after column construction by means of earth pressure coefficient K distribution with depth and radius. Before column construction, there is a thin crust and the earth pressure coefficient shows values of around 0.5 below a depth of 1.0 m. After column construction, the earth pressure coefficient increases to values between 2.0 to 2.5 up to a radius of 2.0 m. Passive failure is nearly reached at the surface. The strong increase of the horizontal stresses during column construction indicates an increase of stiffness in the soil. Adapting the coefficient of earth pressure is a way of modeling the installation procedure when the column penetration is not modeled individually.

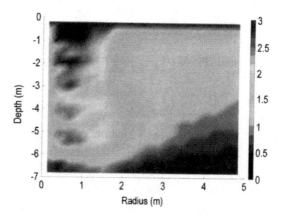

FIG. 8 *Distribution of earth pressure coefficient after column construction, column axis at 0 m radius, column radius at 0.3 m (Weber, 2008)*

A comparison between the behavior of models with and without simulation of column installation is shown Fig. 9a. First it can be seen that there is a strong load dependency. With increasing load, the factor of settlement reduction n ($s_{unimproved}/s_{improved}$) decreases. Second, the model with simulation of column installation behaves stiffer as the model without simulation of column installation until a certain load is reached. At higher load levels, the two models show similar behavior.

A further analysis is given in the Fig. 9b, where the stress concentration *m* in the stone column is shown over column depth in dependency of the load. The stress concentration *m* is defined as the ratio of the vertical effective stress σ'_s in the column over the vertical effective stress σ'_t in the clay at the boundary of the unit cell. It can be seen that the stress concentration varies with depth and load level. The stress concentration is highest at low load levels in a depth of around 2 m, in the calculation reaching values up to 9. At higher load levels, the stress concentration is approximately constant over depth with values around 3. A variation of area reinforcement ratios between 5% and 20% gives similar results. However, the stress concentration of the unit cell model without the simulation of column installation shows values between 2 and 3 for all load levels, indicating no load and depth dependency of the stress concentration.

Fig. 10 shows one selected final results of the plane strain analysis in a superimposed perspective, the deformation of the numerical model and the deformed soil model after centrifuge testing.

Using the 2D numerical model, the results from centrifuge tests are simulated effectively at working load level. Due to the geometrical distortion of the 2D model, the stress state is not reproduced exactly as it would develop in the field and diverge particularly at higher load levels. The stiffness of the column and soil must be modified for 2D analysis. In order to model accurate

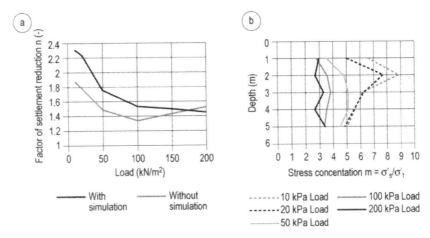

FIG. 9 *Results of axis symmetric numerical model, a) effect of the simulation of stone column installation on the factor of settlement reduction at end of consolidation, b) Stress concentration* m *in stone columns dependent on load and depth,* σ'_s - *vertical effective stress in stone columns,* σ'_t - *vertical effective stress in the clay (Weber, 2008)*

FIG. 10 *Settlement trough and column deformation of numerical and centrifuge model (Weber et al., 2008)*

time dependent behavior, a smear zone has to be introduced, since it has a dominating effect on the consolidation behavior. However, under stress states approaching failure in the soft soil, results from the 2D analysis will not be modeled correctly and the strength of the foundation will be overestimated.

4 CONCLUSIONS

The stone column installation tool referring to a dry bottom feed displacement method was successfully applied in a centrifuge model test to construct stone columns in-flight in a drum centrifuge. The measurement of excess pore water pressure and settlements of an embankment give an indication of the change in the stress state around a stone column during construction and the system behavior of improved ground under load respectively. The results of the centrifuge tests are used for calibration and back analysis with finite element numerical models in order to gain a deeper understanding of processes and interaction within the improved ground.

The change of clay micro-structure and properties during stone column installation could be documented by post test model investigation. By the means of Environmental Scanning Electron Microscopy and Mercury Intrusion Porosimetry, the realignment and orientation of clay particles in the vicinity of the column and the increase of clay density around stone columns was identified. The influence of installation effects on clay density was measured to a distance from column axis of 2.5 times columns radius.

Using an axis-symmetric plane strain finite element model, the installation of stone columns was simulated by a stepwise opening of a cavity within the clay matrix. After column installation, an increase of horizontal stresses was identified leading to earth pressure coefficients of around 2. The application of increased earth pressure coefficient in numerical models is a simplified approach taking column installation effects into account. The stress concentration in the columns is dependent on the depth and the load. Also the load dependency of factor of settlement reduction could be identified particularly for smaller load levels. Without the simulation of stone column installation, the improved ground behaves softer and the stress concentration in the columns is in this case not dependent on depth and load.

Using the 2D numerical model, the results from centrifuge tests are simulated effectively at working load level. Due to the geometrical distortion of the 2D model, soil parameters need to be adapted accordingly. A smear zone has to be introduced for time dependent consolidation analysis. The stress state is not reproduced exactly as it would develop in the field and diverge particularly at higher load levels.

ACKNOWLEDGEMENTS

The author wants to thank the Marie Curie Research Training Network on "Advanced Modelling of Ground Improvement on Soft Soils" and the Swiss National Research Foundation for funding and E. Bleiker, H. Buschor, R. Chikatamarla, A. Ehrbar, M. Gäb, M. Iten, J. Laue, G. Peschke, M. Plötze, A. Privitello, H. F. Schweiger, S. M. Springman and A. Zweidler for their respective contributions to the presented research project.

REFERENCES

Barksdale R. D. and Bachus R.C. (1983). *Design and construction of stone columns*. Volume I., II. Atlanta, Georgia: School of Civil Engineering Georgia Institute of Technology.

Brinkgreve, R. B. J. (2002). *Plaxis 2D - Version 8*. Balkema Publishers, Lisse.

Hird, C. C., Pyrah, I. C. and Russell, D. (1992). Finite element modelling of vertical drains beneath embankments on soft ground. *Géotechnique.* 42(3): 499-511.

Indraratna, B. and Redana, I. W. (1997). Plane-Strain Modelling of Smear Effects Associated with Vertical Drains. *Journal of Geotechnical and Geoenvironmental Engineering (ASCE).* 123(5): 474-478.

Springman, S. M., Laue, J., Boyle, R., White, J. and Zweidler, A. (2001). The ETH Zurich Geotechnical Drum Centrifuge. *International Journal of Physical Modelling in Geotechnics* 1(1): 59-70.

Weber, T. M., Laue, J. and Springman, S. M. (2006). Centrifuge modelling of sand compaction piles in soft clay under embankment load. *VI International Conference on Physical Modelling in Geotechnics.* (eds Ng et al.) Hong Kong. (1): 603-608.

Weber, T. M. (2008). *Modellierung der Baugrundverbesserung mit Schottersäulen.* Veröffentlichungen des Institutes für Geotechnik, ETH Zürich, Band 232.

Weber, T. M., Plötze, M, Laue, J., Peschke, G. and Springman, S. M. (2010). Smear zone identification and soil properties around stone columns constructed in-flight in centrifuge model tests. *Géotechnique* 60 (accepted for publication).

Soft soil Improvement – The Vibro Replacement Techniques

Wehr, J., Wegner, R.
Keller Holding GmbH, Offenbach, Germany

Felix dos Santos, M.
Keller Engenharia Geotécnica Ltda, Rio de Janeiro, Brazil

KEYWORDS soil improvement, vibro replacement, vibro stone columns

ABSTRACT Soil improvement by deep vibro techniques is increasingly used all over the world: vibro compaction to compact loose sands and gravels, vibro replacement stone columns to improve soft and very soft soils as silts and clays. Finally, vibro concrete columns are applied instead of piles to improve all kinds of soil, which can be even of organic nature. An overview of the vibro replacement process and associated application techniques is given together with case histories.

1 INTRODUCTION AND HISTORICAL DEVELOPMENT

For over sixty years, depth vibrators have been used to improve the bearing capacity and settlement characteristics of weak soils. Vibro compaction is probably the oldest dynamic deep compaction method in existence. It was introduced and developed to maturity by the Johann Keller Company in 1936, which enabled the compaction of non-cohesive soils to be performed.

To overcome the limitations of the vibro compaction method with regard to the improvement of cohesive soils, coarse aggregate or gravel was added to the compaction point since approx. 1956 forming a highly compacted stone column surrounded by soil material of enhanced density.

In the beginning of the 1970s, a technique to insert the vibrator into the soil without the aid of simultaneous water flushing was developed. This vibro replacement procedure came to be known as the so-called dry method. Bottom-feed vibrators, which introduce the stones through the vibrator tip during lift, are used avoiding the use of water flushing.

Later it was also possible to install injected stone columns in 1976 by means of an injection of a cement-bentonite suspension near the bottom of the vibrator for special applications. Finally, vibro concrete columns were developed using a conventional concrete pump to deliver the concrete to the bottom of the vibrator via the tremie system.

Probably the oldest recommendation on the use of vibro was issued by the German transport research society in 1979. Later the U.S. department of transportation published the 'design and construction of stone columns manual' (USDT, 1983) followed by the British ICE 'specification for ground treatment' (ICE, 1987) and the BRE publication 'specifying vibro stone columns'

(BRE, 2000). The latest effort has been made by the European community to standardize the execution of vibro works in 'ground treatment by deep vibration' (European Standard EN 14731, 2005).

2   VIBRO REPLACEMENT PROCESSES

The basic principle behind the vibro replacement technique is that particles of non-cohesive soils can be rearranged by means of vibration as with the vibro compaction technique. The ranges of application of the two systems and their limits are shown in Fig. 3.

The essential equipment for deep vibro-techniques is a vibrator – a long heavy tube enclosing eccentric weights, driven by an electric motor.

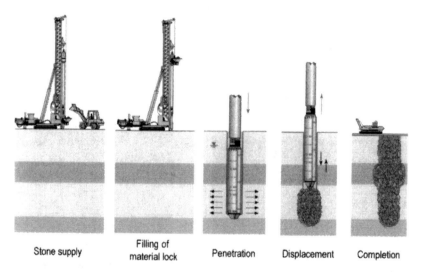

| Stone supply | Filling of material lock | Penetration | Displacement | Completion |

FIG. 1 *Vibro replacement method operating phases (by courtesy of Keller Group)*

The classical method to construct vibro stone columns in cohesive soils involves the use of a strong water jet that ejects water under high pressure from the vibrator tip. The annular space around the vibrator is stabilised by the hydrostatic pressure of the flushing water, which rises to the surface and flushes out loosened soil. The created space is gradually backfilled, through the annulus, with coarse stone fill, which is added at the surface with a wheel loader, slides down along the extension pipes and surrounds the vibrator tip. A highly compacted stone column is formed at the level of vibrator tip by the horizontal movement of the vibrator. This process is known as the conventional, wet vibro replacement method.

More in use now-a-days is the bottom-feed system with carrier equipment, which typically consists of specially designed crawler mounted machines with vertical leaders, known as vibro-cats (Fig. 1). They control the complex bottom-feed vibrators, equipped with material lock and storage units, which deliver fill material to the vibrator tip by means of specialized mechanical or pneumatic feeding devices.

For the installation of vibro concrete columns the tremie system is connected to a mobile concrete pump.

3 VIBRO PLANT AND EQUIPMENT

The equipment developed for the vibro compaction as well as for the vibro replacement processes comprises four basic elements:

- the vibrator, which is elastically suspended from extension tubes and is equipped with air or water jetting systems;
- the crane or base carrier, which supports the vibrator and extension tubes;
- the stone delivery system used for vibro replacement;
- the control and verification devices.

The main piece of equipment used to achieve compaction is the depth vibrator (Fig. 2). The thickness of soil depths to be treated determines the overall length of vibrator, extension tubes and lifting equipment, which, in turn, determines the size of crane or purpose built crawler mounted leaders (vibro-cats).

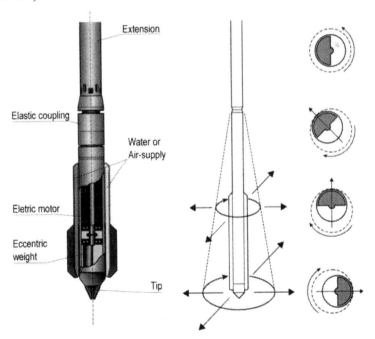

FIG. 2 *Depth vibrator and principle of deep vibro techniques (by courtesy of Keller Group)*

4 DESIGN

Due to their cohesive character, fine soil particles do not separate when exposed to vibrations. This is why for the vibro replacement process coarse grained material consisting of gravel or crushed aggregate is used for backfilling. The feasibility of the technique depends mainly on the grain size distribution of the soil. The range of soil types treatable by the vibro replacement are given in Fig. 3. The degree of improvement achieved by the vibro replacement process depends on the soil being treated, diameter and spacing of stone columns.

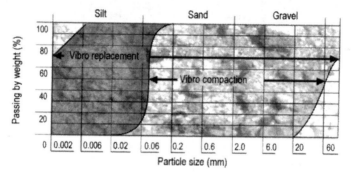

FIG. 3 *Range of soil types treatable by vibro compaction and vibro replacement (stone columns)*

The reduction of compressibility and consolidation time as well as the increase of bearing capacity load and shear strength are the intended effects of the vibro replacement process in soft fine-grained, cohesive soils. Aside from the reduction of overall settlement, the increase in consolidation rate (generated by the stone column's drainage effect) is the target of the vibro column installation.

Priebe's (1995) design method for vibro replacement stone columns has gained common acceptance as a valid method (Fig. 4).

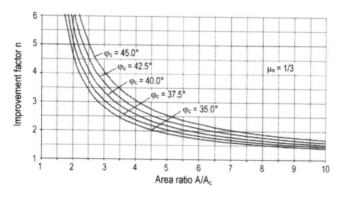

FIG. 4 *Design diagram for improving the ground by vibro replacement stone columns (acc. to Priebe 1995)*

Thus, in Fig. 4, the improvement factor (n = settlement of the unimproved soil/settlement of the improved soil) dependent on angle of internal friction of the stone column material, is related to the ratio of the stone column cross-section and the representative area being treated by the column.

Practical design charts that consider load distribution as well as reduced lateral support on columns situated underneath footing edges have been presented by Priebe (1995). These charts allow the estimation of settlement for an infinite raft supported by an infinite grid of columns as well as for a rigid foundation on a limited number of stone columns.

The time dependent behaviour of stone columns may be analysed using charts proposed by Balaam and Booker (1981). The rates of consolidation are presented in charts as function of the diameter ratio of the unit cell and the column d_e/d, the stiffness ratios of the column and the soil under drained conditions E_1/E_2 and a Poisson's ratio of 0.3 which is assumed equal for the soil and column.

Vibro mortar columns (VMC) and vibro concrete columns (VCC) are ideal for weak, organic soils such as peats and soft clays overlying competent founding strata such as sand, gravels and soft rock. The "bulb end" and frictional components of the VCC's enable high safe working loads to be developed at shallower depths than alternative piling systems. The Priebe model to design vibro replacement was extended to allow also for such stiff columns (Priebe, 2003).

5   APPLICATIONS

Various concepts for creating stone column layouts have been developed. For the support of individual or strip foundations, small groups of columns are employed. Large column grids are placed beneath rigid foundation slabs or load configurations that exhibit flexibility, such as the case with storage tanks and embankments. Due to inherent higher shear resistance, vibro replacement columns are a good choice for the enhancement of slope stability.

6   MONITORING AND TESTING

A major part of the state of the art is to monitor and record the operating parameters of any deep vibro work for quality control purposes.

For the vibro replacement method, the essential parameters of the production process (depth, vibrator energy, feed, contact pressure and stone/concrete consumption) are recorded continuously as a function of time, providing the user with visible and under control data on producing a continuous stone column. Such instrumentation is available and has been used in Europe since the 1980s (Slocombe and Moseley, 1991). A typical print out for stone column construction is given in Fig. 5. Additionally to the online control, the final site records include the position and elevation of columns, the source, type and quality of imported material.

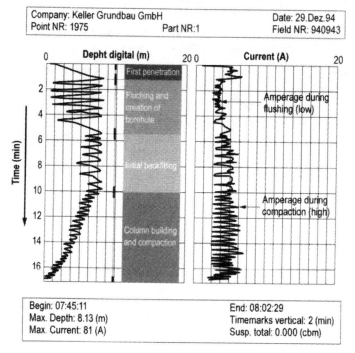

FIG. 5 *Typical installation protocol (by courtesy of Keller)*

Cone penetration tests (CPT) and standard penetration tests (SPT) are commonly used in connection with vibro replacement to analyse the pre-treatment compressibility of the soil and to provide soil data for the layout design. The performance of vibro stone columns is monitored only for large projects using zone load tests. They should be carried out by loading a large area of treated ground, usually by constructing and loading a full size foundation or placing earth fill to simulate widespread loads.

7   CASE HISTORIES

Three case histories, selected from recently completed projects in India, Germany and Australia are presented to illustrate current applications of deep vibro techniques, focusing mainly on foundation for industrial and infrastructure projects.

In case 1 silty sand with fines > 15% have been improved in the provinces Surat/Gujarat in India in 2002. The client was SHELL India. Two LNG-tanks with a diameter of 84 m and a height of 35 m were foreseen in the town of Hazira. The foundation pressure was 340 kN/m². The scope of works was 48,000 linear m of stone columns installed in a square grid of 2.4 m with a length of 16 m each. For quality control reasons 36 numbers of CPT were executed.

In case 2, medium dense, silty, fine to medium graded sands were improved underneath a future railway embankment for high speed railway line in Germany, see Fig. 6. The client was Deutsche Bahn AG – PGS in Berlin and the project was located at Wittenberge/Dergentin on the line Hamburg – Berlin. The scope of works were 6.5 km of railway embankment with 58,000 m of vibro stone columns with a diameter of 0.80 m yielding 9,400 numbers in total with a length of 5.0 m – 7.0 m each. The project was completed in only 7 months.

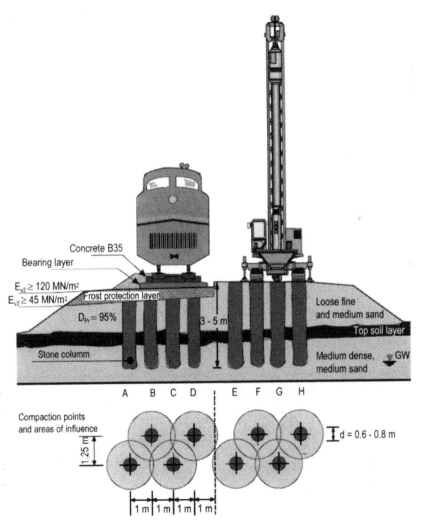

FIG. 6 *Stone column installation along railway embankment*

In case 3, very dense sandy fill underlain by a 2.0 m to 4.5 m thick layer of very soft clay with cone resistances of less than 1 MPa over a lower dense sand profile was improved. The Kooragang Coal Terminal Stockyard Expansion Project in Australia for the Port Waratah Coal Services Ltd., Port of Newcastle, NSW was completed in 2006 having a scope of work of 220,000 m of stone columns. The client's requirement were: A) total post construction settlement of 250 mm for the coal pad area excluding an assumed 100 mm contribution from the deep clay layer; B) vertical differential settlement on the rail track of ±30 mm over a 10 m chord length and ±50 mm over a 25 m chord length and C) minimum requirement of safety against sliding of 1.5 for static loading and 1.1 for earthquake loading. The stability design was made with factors of safety against sliding which varied between 2.35 and 2.75 for static loading and between 1.19 and 1.32 for earthquake loading. All requirements were fulfilled.

8 CONCLUSIONS

Soft soil improvement using deep vibro techniques is applied all over the world. It has been pointed out that vibro stone columns and vibro concrete columns are an economical alternative to piles. Three case histories have been presented to show the applicability to a wide range of soils like loose sands, silty soils and soft clays.

REFERENCES

Balaam, N. P., Booker J. R. (1981). Analysis of rafts and granular piles, International Journal for numerical and analytical methods in Geomechanics, Vol.5, 379-403.

British Research Establishment, BRE (2000). Specifying vibro stone columns.

European Standard. EN14731 (2005). Ground treatment by deep vibration.

Institution of Civil Engineers, ICE (1987) Specification for ground treatment, Thomas Telford.

Priebe, H. J. (1995). The design of vibro replacement, Ground Engineering, December.

Priebe, H. J. (2003) Zur Bemessung von Rüttelstopfverdichtungen, Bautechnik.

Slocombe, B. C., Moseley M. P. (1991). The testing and instrumentation of stone columns, ASTM STP 1089.

U.S. Department of Transportation, USDT (1983) Design and construction of stone columns, Vol. 1.

Section 3 | Monitoring & Performance

Keynote lectures

Overview of Brazilian construction practice over soft soils

Almeida, M. S. S.
COPPE/Federal University of Rio de Janeiro

Marques, M. E. S.
Military Institute of Engineering

Lima, B. T.
COPPE/Federal University of Rio de Janeiro

KEYWORDS embankments over soft soils, site investigation, case histories, construction techniques, monitoring, field performance

ABSTRACT This paper presents an overview of Brazilian experience with construction techniques on soft clays. These are often thick fluvio-marine deposits found along the Brazilian coast, where there is a higher concentration of population and industry. The most used site investigation techniques are summarized, some important case histories are presented and the main construction techniques used in Brazilian practice and their performance are briefly described.

1 INTRODUCTION

About half of the population in Brazil lives in a 100 km strip along the coast, where ports and industries are also concentrated. With increasing development, lowlands areas are increasingly occupied, where soft clay deposits may reach a thickness up to 40 m.

Soft clay engineering started in Brazil about sixty years ago with the birth of geotechnical engineering in the country and since then several studies have been developed along the Brazilian coastal deposits (Pacheco Silva, 1953; Lacerda et al., 1977; Pinto, 1994; Massad, 1994; Almeida and Marques, 2003; Coutinho and Belo, 2005).

This paper presents an overview of the main techniques used in Brazil for construction on these very soft fluvio-marine deposits. Design and calculation methods adopted in Brazilian practice are beyond the scope of the paper and may be found elsewhere (e.g. Almeida, 1996; Schnaid et al., 2001; Massad, 2009).

2 TECHNIQUES FOR SITE INVESTIGATION

Brazilian soft clay deposits are generally quite compressible, with a compression ratio ($C_c/(1+e_0)$) typically around 0.50. The top clay layer usually presents very low strength (design strength can be as low as 3 kPa) and very high water content (w_0), which may reach values of 900% in the top layers (Almeida et al., 2008c).

Undrained strengths (S_u) are mainly obtained by vane tests (using vane borer equipment with shoe protection) but S_u is also obtained from stress history equations (based on the overconsolidation ratio OCR or preconsolidation stress σ'_{vm}). The design strength is obtained using the Bjerrum (1973) correction factor μ, which is usually in the range 0.70 – 0.60 due to the high plasticity of most Brazilian soft clays.

Table 1 presents the main site investigation techniques carried out over soft soils by Brazilian practitioners. Dilatometer (DMT), pressiometer (PMT) and T-bar tests are less routinely carried out.

The Brazilian Code for sampling of soft soils (ABNT, 1997) allows the use of a 75 mm sampler in exceptional situations, but recommends the use of a 100 mm diameter sampler which is mostly used in practice. The sample preparation technique proposed by Ladd and De Groot (2003) for laboratory extrusion has been used in more recent jobs. Sample quality has been evaluated by the method of Lunne et al. (1997) adapted to Brazilian clays (Coutinho, 2007). As it is not easy to obtain good undisturbed samples of the very soft Brazilian clay, *in situ* tests (Danziger and Schnaid, 2000; Coutinho, 2008) are increasingly used.

TABLE 1 TESTS AND GEOTECHNICAL PARAMETERS (ADAPTED FROM ALMEIDA, 1996)

	Test	Main objectives	Parameters (*)
Laboratory	Consolidation	Settlement magnitude and Settlement × time	C_c, C_s, σ'_{vm}, c_v
	Triaxial CU	Stability analysis and modulus for 2D FE analys	S_u, c', ϕ', E_u
Field	SPT	Stratigraphy	w_0 (**)
	Vane	Stability analysis	S_u, S_t, (OCR)
	Piezocone - CPTu	Stratigraphy, Settlement × time from dissipation tests	S_u profile, c_h (OCR; K_0; E_{oed})

(*) other parameters that could be obtained from test are given between brackets; (**) w_0 is determined from the soil collected at the tip of the SPT sampler and could be used in correlations.

3 SOME IMPORTANT BRAZILIAN CASE HISTORIES

Some well-instrumented and analyzed case histories in Brazil executed during the period 1974–1994 were very useful in forming the basis of soft clay engineering practiced in the country in subsequent years.

Three cases are selected and briefly described below: the Sarapuí Test Embankment I (Ortigão et al., 1983), Juturnaíba Test Embankment (Coutinho and Lacerda, 1989) and the Sergipe Breakwater (Brugger et al., 1998). The latter was built in stages during four years but the Sarapuí and Juturnaíba embankments were taken to failure in about one month's time. This construction time is usually close enough to an undrained condition in practical terms as far as stability analysis is concerned. However, this may not be the case as far as deformation analysis is concerned, as the results to be presented below illustrate.

The Sarapuí Embankment I (Ortigão et al., 1983) was built until rupture, 30 days after beginning of loading. Subsequent stability studies of this test embankment (Almeida, 1985; Sandroni, 1993) indicated that the actual failure took place when the embankment reached a height of 2.5 m. Almeida and Marques (2003) have described extensive comparative analysis between laboratory and *in situ* test results and field observations.

Back analysis of Embankment I rupture using a 3D failure surface has shown that the correction factor for Sarapuí clay is $\mu = 0.70$ (Sandroni, 1993; Pinto, 1992). This value lies slightly above (for a typical IP = 80) the μ versus IP relationship proposed by Azzouz et al. (1983), for this type of failure, as shown in Fig. 1, where other Brazilian case histories are also included, most of which lie on the vane strength correction curves, with the exception of the Juturnaíba test embankment.

Embankment I was analyzed numerically (Almeida and Ortigão, 1982) using the modified Cam-clay model. Analyses were performed for both undrained and partially drained conditions (using Biot 2D consolidation

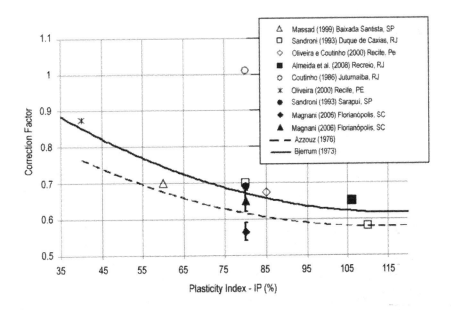

FIG. 1 *Vane strength correction versus plasticity index Brazilian case histories*

theory). Good overall agreement between measured and numerical values of base settlements and pore pressures was observed for the partially drained case, but not for the undrained condition, as shown in Fig. 2. Therefore, it seems that even for a one month long embankment loading some drainage may have taken place.

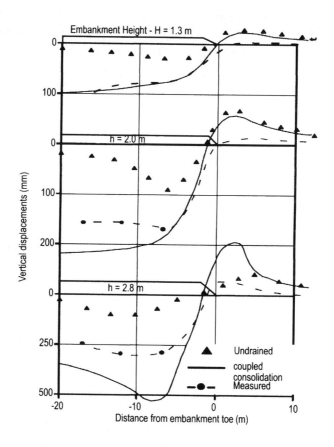

FIG. 2 *Settlements at the embankment I base at various stages of construction*

The Juturnaíba test embankment was another important case history (Coutinho and Lacerda, 1989). This test embankment was brought to failure with the aim of gaining a better understanding of the foundation behaviour of the Juturnaíba dam. The local subsoil consisted of a 7.5 m thick soft clay layer, composed of six well studied sub-layers. The Juturnaíba test embankment was modelled numerically (Antunes, 1996) using the modified Cam-clay model with 2D Biot consolidation for the soft clay layers. Fig. 3 shows measured and predicted displacements for two embankment heights where good agreement is observed. Good overall agreement between measurements and predictions was also obtained for horizontal displacements and pore pressures.

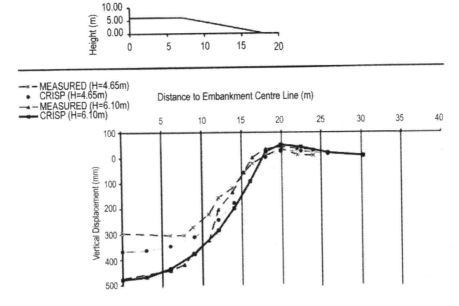

FIG. 3 *Computed and measured vertical displacements at the base of the Juturnaíba embankment (Antunes, 1996)*

The Sergipe Breakwater is located 2.5 km from the Northeast coast and has a total length of 800 m and height of 15.25 m (10 m underwater). The embankment was built in 5 stages, until the desired elevation was reached. The geotechnical investigations carried out showed the presence of about 4 m of sand over a 7 m thick soft clay layer. The geotechnical characteristics of the local soft clay were summarized by Sandroni et al. (1997). Brugger et al. (1998) presented a numerical analysis of the construction of the breakwater using modified Cam-Clay type computer models for the clay layers.

Fig. 4 shows the location of the extensometer (SD-03), installed before the 4th stage construction, along with a comparison of the instrument readings with numerical analysis results. The authors noted that the settlements predicted for the 5th stage are a little higher than those measured, which is a situation that was reversed after 1,300 days of reading, and the largest settlements in the field (final expected value of 1.45 m) can be explained by secondary compression, which was not predicted in the numerical model.

The three cases described above used more conventional construction techniques, with no special requirements regarding stability or settlement control such as reinforcement or vertical drains. Thus, they could be more easily analyzed and set the basis for more elaborate construction techniques used subsequently in the country in softer and more compressible clay deposits.

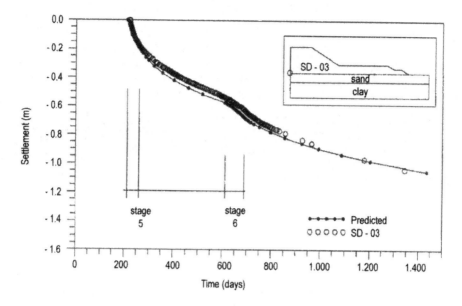

FIG. 4 *Predicted and measured settlements at Sergipe breakwater (Brugger et al., 1998)*

4　CONSTRUCTION TECHNIQUES

The construction techniques mostly used nowadays to build embankments over soft clays in Brazil are outlined in Table 2. Some of these techniques will be discussed in greater detail in this paper. The methodology mostly used in relation to Brazilian very soft soils combines the use of stage construction, berms, geogrid reinforcement and vertical drains under the embankments.

Although these more conventional techniques have been in Brazilian practice for decades, other construction techniques have been introduced more recently in Brazil. Examples of these are piled embankments on geogrid platforms, use of granular columns (encased or not) and lightweight fills which are described in some length in the present paper. Studies are also under way in some Brazilian projects to use other solutions in the near future such as vacuum preloading.

Jet-grouting and shallow mass stabilization (Stabtec® technique) have also been used in the country in various soft clay projects but is beyond the scope of this paper to discuss these techniques. Other techniques such as lime columns, deep mixing and dynamic compaction have not yet been used in the country, to the authors' knowledge.

4.1　Vertical drains

Vertical drain has been the most widely used technique for soil improvement in Brazil for some time. The accumulated experience in Brazil and elsewhere

TABLE 2 CONSTRUCTION METHODS

Method	Comments	Brazilian studies
Soft Soil Removal	Environmental impact on neighbouring areas are the main concern.	Vargas (1973), Cunha and Wolle (1984)
Soft soil expulsion by controlled failure	Often used for shallow deposits; method highly dependent on local experience; the remaining soft soil left should be assessed by borings.	Zaeyen et al. (2003)
Stage construction	Mostly used with vertical drains; requires careful design and control (increase in clay strength); unfavorable for tight schedules.	Almeida et al. (1985) (*), Almeida et al. (2008)
Vertical drains and/or fill surcharge	Used to accelerate settlements; wide accumulated experience. Temporary surcharge useful to decrease/suppress secondary settlements.	Almeida et al. (2001), Sandroni and Bedeschi (2008)
Equilibrium berms and/or reinforcement	Quite commonly adopted; necessary to assess whether specified tensile force is actually mobilized in situ.	Palmeira and Fahel (2000), Magnani et al. (2009)
Light fill materials	Favorable for tight schedules, relatively high cost; its use has increased recently.	Sandroni (2006)
Piled embankments	Favorable for tight schedules, may present different layouts and materials, for very small settlements.	Almeida et al. (2008b), Sandroni and Deotti (2008)
Granular columns (Granular piles)	Granular columns may be encased or not with geotextiles; settlement acceleration owing to draining nature of granular columns; geogrids are sometimes placed above the granular piles.	Almeida et al. (1985), Mello et al. (2008), Garga and Medeiros (1995)
Vacuum preloading	May partially substitute fill surcharge; horizontal displacements much lower than for standard surcharge.	Marques and Leroueil (2005)(*)

(*) Studies carried out by Brazilian researchers on clay deposits of other countries.

(Almeida and Ferreira, 1992; Sandroni, 2006; Saye, 2001) indicates that the efficiency of vertical drains depends on the distance between drains. Nowadays it is common practice in Brazil to limit the minimum distance between prefabricated drains to 1.5 m in order to avoid remoulding of the clay, although the mandrel and plate anchor dimensions are also factors to be considered in respect of remoulding and minimum drain distance.

A settlement study (Almeida et al., 2001) was performed on an embankment built on Barra da Tijuca highly compressible organic clay, in which water content varies from 500% near the surface to 100% at the base of the clay deposit. The 0.60 m thick drainage blanket provided support for the equipment used to install a prefabricated vertical drain (PVD) in a 1.70 m triangular grid. The settlements covered a period up to 2 years, when the average degree of consolidation reached around 90%, and at that time the average gain in strength of the whole clay layer was close to $\Delta S_u / \Delta \sigma'_v = 0.22$. Values of c_h were measured and the generally good agreement obtained in dissipation tests, using a piezocone to compare with field values, using Asaokas Method (1978) modified by Magnan and Deroy (1980), suggests that during the period of analysis secondary consolidation was negligible compared to primary consolidation, as shown in Table 3.

TABLE 3 SUMARY OF A C_h VALUES OF BARRA DA TIJUCA CLAY (ALMEIDA ET AL., 2001)

Method	Range of c_h variation (10^{-8} m^2/s)	c_h (average)
Asaoka (1978)	3.7 – 10.5	6.8
Piezocone	2.4 – 13.7	8.2
Special radial oedometer tests	3.6 – 6.8	5.0

Fig. 5 shows the settlements at the centre of an embankment constructed over a 1.4 m triangular grid of PVD, at Recreio site, Rio de Janeiro city (Almeida et al., 2008a). In the frontal region of the site, where the PVD were installed, the soft clay thickness varied from 4 to 11 m and due to its high

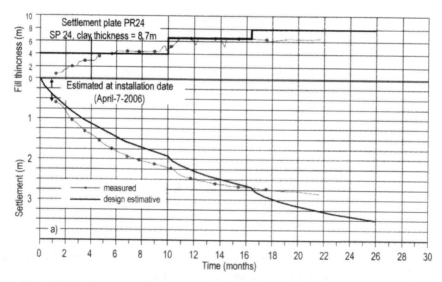

FIG. 5 *Typical settlement plate data of Recreio embankment (Almeida et al., 2008)*

compressibility, in order to stabilize a 3 m high embankment, it was necessary to build a 7 m thick embankment in three stages, over vertical drains and reinforcement at the borders of the embankment. Due to the high compressibility and low undrained strength of Brazilian soft clays, and also the high values of secondary settlements, the use of PVD associated with surcharge can be very expensive due to fill volumes, reinforcement and the long construction schedules required. All these factors have led to the increasing use of alternatives construction techniques.

4.2 Reinforced embankments

The use of reinforcement at the base of embankments on soft clays started in Brazil in the 1980s (e.g. Ortigão and Palmeira, 1982) and since then has increased rapidly. In the early years, lower strength unwoven geotextiles were more widely used. However, in the last fifteen years higher strength geosynthetics, woven geotextiles and geogrids have become available and have gained wider acceptance. A number of case histories of reinforced embankments in Brazil have been reported and some of these are briefly described below.

Palmeira and Fahel (2000) presented the performances of four geosynthetic reinforcement highway abutments built over soft foundations, including a geogrid reinforced bridge at BR-101 in south Brazil. In this case the new embankment was built with part of its width resting on the old embankment, as shown in Fig. 6. The instrumentation showed that the reinforcement was very effective at reducing lateral movement of the soft foundation soil. In spite of the large deformations imposed on the structures, the overall performance of this embankment was satisfactory, mainly because of the flexible nature of the geosynthetic reinforcement used.

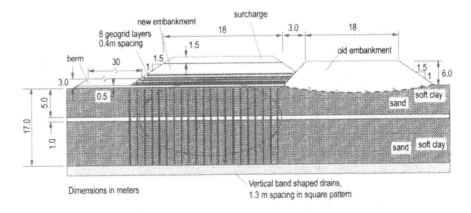

FIG. 6 *Cross-section along the transverse direction of the embankment (Palmeira and Fahel, 2000)*

Fahel and Palmeira (2002) described another case history of reinforced bridge abutment which almost collapsed before its planned final height had been reached. Two main reasons for the instability of the embankment were identified in this case: the use of a uniaxial geogrid with quite different tensile strength values along the longitudinal and transverse directions and the rapid rate of construction.

Magnani et al. (2009, 2010a) reported the construction of two reinforced test embankments built to failure with measurement of the tensile forces applied to the geotextile reinforcement. The performances of these embankments are summarized in a paper presented to this symposium (Magnani et al., 2010b), so further information is not presented here.

4.3 Embankments on granular columns

Techniques used for granular column installation (e.g. Garga and Medeiros, 1995) did not present adequate performance in Brazil in the last century. This was mainly due to issues such as type of equipment used and selection of granular column diameter and spacing. Modern equipment and techniques have been recently introduced and important soft clay improvement works have been carried out in Brazil using granular columns.

One of these works was the construction of an iron ore stockyard in the CSA ThyssenKrupp Steel Plant, covering an area of approximately $9 \, km^2$, located 50 km from Rio de Janeiro city. Due to vertical stresses equal to 340 kPa imposed by the 13 m high iron ore stockpiles, vibro-replacement techniques with dry and wet methods were used. Subsoil conditions at the site consist of fluvial and fluvio-marine sediments of quaternary origin, alternating stratification of sands, silts and clays, as well as fluvial gravel and younger mangrove deposits (Wegner et al., 2009 and Marques et al., 2008). The compressible layers are 12 to 15 m deep, reaching 17 m in some places. It was estimated that, without the gravel columns, the settlements would reach values around 1.5 m. Fig. 7 shows the geometry and geotechnical model used for the solution with gravel columns.

The gravel columns were installed at intervals between 1.75 m and 2.20 m in a square arrangement, with an average length of 12 m and a total length of approximately 400,000 m. A load transfer platform was constructed above the gravel columns with the use of two layers of high-strength bi-directional geogrids. Settlement rates from 2 to 3 cm per month were observed in the iron ore stockyard, but decreased after a period of approximately 4 months (Wegner et al., 2009).

Other stone column works carried out in Brazil recently (Felix, 2009) include the construction at the BR-101 highway toll station in Araquari, State of Santa Catarina, of 10,305 m of column (diameter 0.80 m; length 7–10 m;

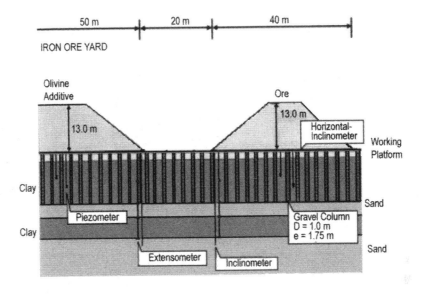

FIG. 7 *Geometry and geotechnical model used for the solution with gravel columns (Wegner et al., 2009)*

spacing 1.80–2.50 m). Also, 5,378 m of column were installed for the enlargement of the Ubá municipal airport, State of Minas Gerais (diameter 0.70 m; length 6–9 m; spacing 3.5–5.0 m).

4.4 Embankments on encased sand columns

Encased sand columns were used for the first time in South America on a highway (Mello et al., 2008) near the city of São José dos Campos, 100 km from São Paulo city. The subsoil at this site is composed of two layers of soft clay separated by a layer of silty sand. Table 4 shows the range of geotechnical parameters of these two clay layers.

The columns were installed using closed end Franki pile equipment. After the Franki tube installation, the sand was deposited within the geosynthetic casing and the tube was removed with the aid of a vibratory hammer.

TABLE 4 RANGE OF GEOTECHNICAL PARAMETERS LABORATORY

Parameter	Range of values
Bulk weight (kN/m^3)	13.2 – 15.4
Vertical consolidation coefficient c_v (x 10^{-8} m^2/s)	1.3 – 50
OCR	1.2 – 1.6
CR = $C_c/(1+e_0)$	0.17 – 0.40
Undrained strength (kPa)	6 – 12

Fig. 8 shows the final stages of implementation of an encased column installation and the column inside. Table 5 summarizes the characteristics of the columns used and some monitoring results.

FIG. 8 *Finished geosynthetic encased sand column*

TABLE 5 SUMMARY OF THE CHARACTERISTICS OF COLUMNS AND RESULTS

Characteristics	Values
Diameter of columns	0.70 m
Geotextile used in the jacking	Final stress of 50 kN/m and rigidity of 1,000 kN/m
Length of columns	10.0 m
Spacing	1.8 and 2.2 m
Settlement measured (max.)	100 mm
Time of stabilization after the beginning of readings	6 months

Encased sand columns were also used in sites of the CSA ThyssenKrupp Steel Plant iron ore stockyards (Alexiew and Moormann, 2009) where the local soil was composed of layers of very soft and compressible clay. The soil and column characteristics are summarized in Table 6.

4.5 Lightweight fill

Different types of lightweight materials may be used in order to reduce settlements and increase the overall embankment stability. Among these materials, expanded polystyrene EPS is nowadays the most widely used as it has the

TABLE 6 RANGE OF GEOTECHNICAL PARAMETERS OF THE LAYERS AND CHARACTERISTICS OF COLUMNS IN THE CSA WORK (ALEXIEW AND MOORMANN, 2009)

Soil parameters	Values
Thickness of clay layers	Up to 20 m of clay layer with 8 m or more compressible clay
Oedometric modulus	$0.2 - 0.5\,\text{MN/m}^2$
c_v	$2 - 4 \times 10^{-8}\,\text{m}^2/\text{s}$
S_u	$5 - 15\,\text{kN/m}^2$

Columns characteristics	Values
Diameter of columns	0.78 m
Length of columns	10 to 12 m
Spacing	$2.0 \times 2.0\,\text{m}$
Geotextile used in the jacking	Ringtrac R 100/250 and 100/275

lowest specific weight and combines high strength and low compressibility. The use of EPS blocks as a light fill material is still not widely used in Brazil, owing to its high cost, but some uses in São Paulo (Gonçalves and Guazzelli, 2004) and in Rio de Janeiro (Sandroni, 2006; Lima and Almeida, 2009) have been reported, as exemplified in Fig. 9.

FIG. 9 *Use of EPS blocks as lightweight fill material: a) cross section showing blocks and protective cover; b) finished pavement (Lima and Almeida, 2009)*

4.6 Piled embankments

Piled embankments with a geogrid or concrete platform can be an alternative with technical, economical and schedule advantages, when compared to a reinforced embankment with berms on vertical drains, mostly due to the high value of fill material necessary for more conventional solutions. For very soft clays like the ones described here, piled embankment solutions for roads may be more economic than embankments on vertical drains for soft layer thicknesses greater than about 12 m (Almeida and Marques, 2004). However, it is important to analyze each case, since stratigraphy, soil parameters and fill costs vary over a wide range as shown by Nascimento (2009) and thus the clay thickness threshold could be lower than 12 m for some deposits.

The performance of piled embankments built over Barra soft clay deposits were described by Almeida et al. (2007, 2008) and Sandroni and Deotti (2008). The prevailing mechanism in these cases was the membrane effect due to the low ratio of embankment height to pile caps span. When the membrane effect prevails, geogrid deformation should be expected in the future, and the pavement should be light and flexible, allowing future settlements and maintenance.

Based on the performance of these piled embankments, it was observed in very soft clays that the working platform causes important settlements, the magnitude of which is not negligible.

Palmeira and Fahel (2000) describe the behavior of the reinforced abutment with the highest reinforced structure (7.3 m), where no settlement was observed due to the efficiency of the piles along the embankment base. However, horizontal displacements of the wall crest of the order of 40 mm (approximately 0.55% of the wall height) were measured.

4.7 Vacuum Consolidation

Brazilian expertise with vacuum consolidation comes from a study performed at a site in Canada (Marques and Leroueil, 2005), in a research program carried out by Laval University and COPPE/UFRJ. Two trial embankments were built in order to study the technique of preconsolidation by vacuum (trial Embankment A) and by vacuum and heating (trial Embankment B). The site had a water table at a depth of 1.5 m, inside the weathered crust, and a flow of water towards the underlying till layer. Results of this experiment are summarized in Fig. 10 where good correspondence is seen between the applied vacuum pressure, vertical strains (measured by special vertical extensometers) and pore pressure.

The vacuum technique is a good option for very large areas, where often the superficial very soft clay has very low undrained strength. For these sites, like most Brazilian deposits, the technique presents two advantages: it is possible to attain high stresses without rupture and there is a significant economy with the fill material generally used as surcharge. Another advantage deals with the logistics of obtaining fill material near the embankment area and the transportation and disposal of the material, which could be of great importance for the construction schedule.

Some studies on the subject are being conducted at present in Brazil to use the vacuum technique at nearby shore sites. The main drawback with the use of vacuum consolidation in Brazilian deposits is the presence of sand layers in soft clay deposits (Massad, 2009) as in these cases it is necessary to use plastic diaphragm walls to maintain system tightness.

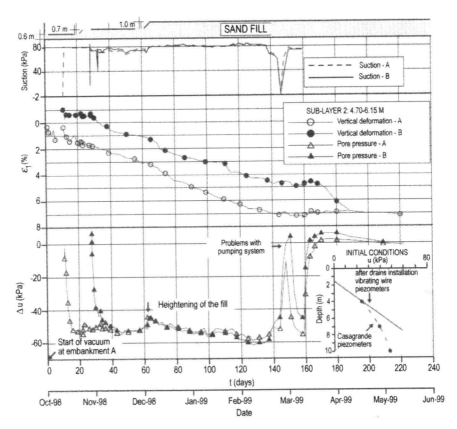

FIG. 10 *Vertical deformation of sub-layers 2A and 2B (fill B) and pore pressure measured in the middle of these layers, at a depth of 5.4 m (Marques and Leroueil, 2005)*

5   INSTRUMENTATION FOR MONITORING STABILITY

The instrumentation for the construction of stage embankments consists mostly of settlements plates and inclinometers, and sometimes piezometers. To evaluate the gain in clay strength between stages, vane tests are carried out. The use of extensometers is generally associated with a more extensive instrumentation programme, generally for important projects or research.

Settlement control is usually based on settlement and piezometer measurements (e.g. Almeida and Ferreira, 1992; Sandroni and Bedeschi, 2008) but a discussion of these topics is beyond our scope.

For stability control based on inclinometer results, Almeida et al. (2000) proposed that, for distortion velocities (v_d) higher than 1.5%/day, there are regions where plasticization occurs and rupture could be imminent, and the interruption of surcharge is recommended. For v_d values between 0.5%/day and 1.5%/day, special care should be taken, but for $v_d \leq 0.5$%/day only the continuous monitoring is required until stabilization.

Sandroni et al. (2004) proposed an empirical method for the evaluation of the safety of embankments over soft soils. The procedure takes into account that the volume due to settlement (V_v) and the volume due to horizontal displacement (V_h), for undrained conditions and plane analysis, should be the same. They proposed that:

- for undrained behavior V_v/V_h are close to one, and non planar undrained behavior $V_v/V_h > 1$;

- for drained behaviour $V_v/V_h > 1$.

During loading V_v/V_h could vary since field behavior is a mix of undrained and drained behavior and deformations are not planar. When close to rupture V_v/V_h tends to 1 very quickly, however when loading is interrupted V_v/V_h increases with time, going to stabilization

Magnani et al. (2008) applied the procedures of Almeida et al. (2000) and Sandroni et al. (2004) to three experimental embankments assumed to rupture, one with drains and reinforcement, another with reinforcement and a third which was conventional (without drains or reinforcement) and obtained satisfactory results for the three embankments.

6   CONCLUSIONS

This paper describes techniques to build embankments on very soft Brazilian clays. The used site investigation techniques involve *in situ* tests (mainly SPT, CPTu, VT) and laboratory tests (oedometer and triaxial tests), which require careful sampling and specimen preparation in order to get reliable parameters for very soft clays. The design strength profile is mainly based on the corrected vane strength supported by normalized undrained strength equations. Coefficient of consolidation design values are usually based on both oedometer tests and CPTu dissipation tests.

The Brazilian case histories of Sarapuí, Juturnaíba and Sergipe breakwater, which were performed in the 1970s to early 1990s, are well-studied and were very useful in understanding the field behavior of Brazilian soft clays.

A number of more elaborate construction techniques involving the use of reinforcement, berms, vertical drains, temporary surcharge and stage construction are then described by means of practical applications. In very soft Brazilian clays, it is common to make combined use of all these techniques in the same project or site. However, the overall construction involved is still extensive due to the nature of these very soft deposits.

Less conventional construction techniques such as piled embankments with geogrid platforms and soil improvement techniques using granular columns and lightweight fills have also been adopted more recently in these very soft deposits. These less conventional techniques are gaining wider acceptance as they allow shorter overall construction times.

ACKNOWLEDGEMENTS

The geotechnical data summarized herewith result from a large number of studies, particularly Master's and PhD dissertations and the authors are very much indebted to all those involved in these studies. The authors would like to thank the Brazilian Research Council (CNPq-Pronex Program) and State of Rio de Janeiro Research Agency (FAPERJ – Cientistas do Estado Program) for their financial support and to Marcos B. Mendonça and José Renato M. Oliveira for their comments in the final version of the paper.

REFERENCES

ABNT. (1997). NBR9820, Assessment of undisturbed low consistency soil samples from borehole (*In Portuguese*). Brazilian Code.

Alexiew, D. and Moormann, C. (2009). Foundation of a coal/coke stockyard on soft soil with geotextile encased columns and horizontal reinforcement. *Proceedings of the 17th ICSMGE*, CD-ROM

Almeida, M. S. S. and Ferreira, C. A. M. (1992). Field, in situ and laboratory consolidation parameters of very soft clay. Predictive Soil Mechanics. *Proceedings of the Wroth Memorial Symposium*, Oxford, pp. 73–93.

Almeida, M. S. S. and Marques, M. E. S. (2003). The behaviour of Sarapuí soft organic clay. In: T. S. Tan, K. K. Phoon, D. W. Hight and S. Leroueil (eds.). *Characterisation and Engineering properties of Natural Soils*, vol. 1, pp. 477–504.

Almeida, M. S. S. and Marques, M. E. S. (2004). Embankments on thick compressible soft clay layers (*In Portuguese*). *2° Congresso Luso-Brasileiro de Geotecnia*, Aveiro. pp. 103–112.

Almeida, M. S. S. and Ortigão J. A. R. (1982). Performance and Finite Element Analyses of a Trial Embankment on Soft Clay. *International Symposium on Numerical Models in Geomechanics*, Zurich. pp. 548–558.

Almeida, M. S. S. (1985). Discussion on: "Embankments Failure on Clay near Rio de Janeiro". *Journal of Geotechnical Engineering*, ASCE, vol. 111 (2), pp. 253–256.

Almeida, M. S. S. (1996). Embankments on soft clays: design and performance (*In Portuguese*). Ed. UFRJ, 215 p.

Almeida, M. S. S., Davies, M. C. R. and Parry, R. H. G. (1985). Centrifuged embankments on strengthened and unstrengthened clay foundations. *Géotechnique*, vol. 35 (4), pp. 425–441.

Almeida, M. S. S., Ehrlich, M., Spotti, A. P. and Marques, M. E. S. (2007). Embankment supported on piles with biaxial geogrids. *Journal of Geotechnical Engineering*, Institution of Civil Engineers, ICE, UK vol. 160 (4), pp. 185–192.

Almeida, M. S. S., Marques, M. E. S, Lima, B. T. and Alvez, F. (2008a). Failure of a reinforced embankment over an extremely soft peat clay layer. *4th European Conference on Geosynthetics-EuroGeo*, 2008, Edinburgh, vol. 1, pp. 1–8.

Almeida, M. S. S., Marques, M. E. S., Almeida, M. C. F. and Mendonça, M. B. (2008b). Performance of two "low" piled embankments with geogrids at Rio de Janeiro. *The First Pan American Geosynthetics Conference & Exhibition*, 2–5 March, Cancun, Mexico, on CD.

Almeida, M. S. S., Marques, M. E. S., Miranda, T. C. and Nascimento, C. M. C. (2008c). Lowland reclamation in urban areas. *Workshop on Geotechnical Infrastructure for Mega Cities and New Capitals*, TCIU - ISSMGE, Búzios, 25–26 August, on CD.

Almeida, M. S. S., Oliveira, J. R. M. S. and Spotti, A. P. (2000). Prediction and performance of embankments over soft soils: stability, settlements and numerical analysis (*In Portuguese*). *Previsão de Desempenho × Comportamento Real*, ABMS, São Paulo, pp. 69–94.

Almeida, M. S. S., Spotti, A. P., Santa Maria, P. E. L., Martins, I. S. M. and Coelho, L. B. M. (2001). Consolidation of a very soft clay with vertical drains. *Geotechnique*, vol. 50 (2), pp. 633–643.

Antunes Filho, V. (1996). Numerical analysis of Juturnaíba embankment on soft organic clay (in Portuguese). MSc. Thesis, COPPE/UFRJ, Rio de Janeiro, Brazil.

Asaoka, A. (1978). Observational Procedure of Settlement Prediction. Soil and Foundations, vol. 18 (4), pp 87-101.

Azzouz, A. S., Baligh, M. M. and Ladd, C. C. (1983). Corrected Field Vane Strength for Embankment Design. *Journal of Geotechnical Engineering*, ASCE, vol. 109 (5), pp. 730–734.

Bjerrum, L. (1973). Problems of soil mechanics and construction on soft clays and structurally unstable soils. *8th International Conference on Soil Mechanics and Foundation Engineering*, Moscow, vol. 3, pp. 111–159.

Brugger, P. J., Almeida, M. S. S., Sandroni, S. S. and Lacerda, W. A. (1998). Numerical analysis of the breakwater construction of Sergipe Harbour, Brazil. *Canadian Geotechnical Journal*, vol. 35 (6), pp. 1018–1031.

Coutinho, R. Q. and Bello, M. I. M. C. (2005). Aterro sobre Solo Mole. In: Alexandre Duarte Gusmão e outros. (Org.). *Geotecnia do Nordeste/ABMS*. 1 ed. Recife: Editora Universitária/UFPE, pp. 111–153.

Coutinho, R. Q. and Lacerda, W. A. (1989). Strength Characteristics of Jurtunaíba Organic Clays. *XII Int. Conf. on Soil Mechanics and Foundation Engineering*, Rio de Janeiro, vol. 03, pp. 1731–1734.

Coutinho, R. Q. (2007). Characterization and engineering properties. *The Second International Workshop on Characterization and Engineering*

Properties of Natural Soils. Editors Tan, Phoon, Higth & Leroueil. Singapore, pp. 2049–2100.

Cunha, M. A. and Wolle, C. M. (1984). Use of aggregates for road fills in the mangrove regions of Brazil. *Bulletin of International Association of Engineering Geology*, Paris, vol. 30, pp. 47–50.

Danziger, F. A. B and Schnaid, F. (2000). Piezocone tests: procedures, recommendations and interpretation *(in Portuguese)*. *Brazilian Seminar of Field Investigation*, São Paulo, Brazil, vol. 3, pp. 1–51.

Fahel, A. R. S. and Palmeira, E. M. (2002). Failure Mechanism of a Geogrid Reinforced Abutment on Soft Soil. *7th International Conference on Geosynthetics*, Nice. Lisse: Balkema. vol. 4, pp. 1565–1568.

Felix, M. (2009). Personal communication

Garga, V. K. and Medeiros, L. V. (1995). Field Performance of the Port of Sepetiba Test Fills. *Canadian Geotechnical Journal*, 32 (1), pp. 106–121.

Gonçalves, H. H. S. and Guazzelli, M. C. (2004). Use of EPS on embankments over soft soils *(in Portuguese)*. *II Congresso Luso-Brasileiro de Geotecnia*, Aveiro, pp. 73–82.

Lacerda, W. A., Costa Filho, L. M., Coutinho, R. Q. and Duarte, A. R. (1977). Consolidation characteristics of Rio de Janeiro soft clay. *Conference on Geotechnical Aspects of Soft Clays*, Bangkok, pp. 231–244.

Ladd, C. C. and DeGroot, D. J. (2003). Recommended Practice for Soft Ground Site Characterization. *12th Pan American Conference on Soil Mechanics and Geotechnical Engineering*, Boston, vol. 1, pp. 3–57.

Lima, B. T. and Almeida, M. S. S. (2009). Use of lightweight EPS fill material in Barra da Tijuca very soft clay *(in Portuguese)*. *Conf. Sudamericana Ingenieros Geotécnicos Jóvenes*, Córdoba - Argentina, vol. 1, pp. 153–156.

Lunne, T. Berre, T. and Strandvik, S. (1997). Sample disturbance effects in soft low plastic Norwegian Clay. *Recent Developments in Soil and Pavement Mechanics*. M. Almeida (ed.), Rio de Janeiro, pp. 81–102, Rotterdam: Balkema.

Magnan, J. P. and Deroy, J. M. (1980). Analyse graphique des tassements observés sous les ouvrages. *Bulletin Liason Laboratoire des Ponts et Chaussés*, 109, p. 9-21.

Magnani, H. O., Almeida, M. S. S. and Ehrlich, M. (2008). Construction Stability Evaluation of Reinforced Embankments over Soft Soils. *1st Pan American Geosynthetics Conference & Exhibition*, Cancum. CD-ROM. pp. 1372–1381.

Magnani, H. O., Almeida, M. S. S., Almeida and Ehrlich, M. (2009). Behaviour of two reinforced test embankments on soft clay. *Geosynthetics International*, ICE, vol. 16 (3), pp. 127–138.

Magnani, H. O., Ehrlich, M. and Almeida, M. S. S. (2010a). Embankments over soft clay deposits: the contribution of basal reinforcement and surface sand layer to stability. *Journal of Geotechnical and Geoenvironmental Engineering*, ASCE, vol. 136 (1), pp. 260–264.

Magnani, H. O., Almeida, M. S. S. and Ehrlich, M. (2010b). Reinforced test embankments on Florianopolis very soft clay. *Symposium on New Techniques for Design and Construction in Soft Clays*, Guarujá, Brazil, May 2010.

Marques, M. E. S. and Leroueil, S. (2005). Preconsolidating clay deposit by vacuum and heating in cold environment. In: Buddhima, I. and Jian, C. (eds), *Ground Improvement-Case Histories*, Elsevier Geo-Engineering Book Series 3, pp. 1045–1063.

Marques, M. E. S., Lima, B. T., Oliveira, J. R. M., Antoniutti Neto, L. and Almeida, M. S. S. (2008). Geotechnical Characterization of a soft soil deposit at Itaguaí, Rio de Janeiro *(in Portuguese)*. *IV Congresso Luso-Brasileiro de Geotecnia*, Coimbra, Portugal.

Massad, F. (1994). Properties of marine sediments. *Solos do Litoral do Estado de São Paulo*. ABMS, pp. 99–128.

Massad, F. (2009). *Soils of Santos lowlands: characteristics and geotechnical properties (in Portuguese)*. Oficina de Textos, 247 pp.

Mello, de L. G., Mandolfo, M., Montez, F., Tsukahara, C. N. and Bilfinger, W. (2008). First Use of geosynthetic encased sand columns in South America. *1st Pan American Geosynthetics Conference & Exhibition*, Cancun, Mexico, 2–5 March 2008, on CD.

Nascimento, C. M. C. (2009). Assessment of alternative construction techniques of embankments of urban roads over soft soils. MSc. Dissertation *(in Portuguese)*, Military Institute of Engineering, Rio de Janeiro, Brazil.

Ortigão, J. A. R. and Palmeira, E. M. (1982). Geotextile performance at an access road on soft ground near Rio de Janeiro. *2nd International Conference on Geotextiles*, Las Vegas, USA, vol. 2, pp. 353–358.

Ortigão, J. A. R., Werneck, M. L. G. and Lacerda, W. A. (1983). Embankment failure on clay near Rio de Janeiro. *Journal of the Geotechnical Engineering Division*, ASCE, vol. 109 (11), pp. 1460–1479.

Pacheco Silva, F. (1953). Shearing strength of a soft clay deposit near Rio de Janeiro. *Géotechnique*, vol. 3, 300–306.

Palmeira, E. M. and Fahel, A. R. S. (2000). Effects of Large Differential Settlements on Embankments on Soft Soils. *2nd European Conference on Geosynthetics-EuroGeo*, Bolònha: Pàtron Editore, vol. 1, pp. 261–267.

Pinto, C. S. (1992). Topics of the contribution of Pacheco Silva and considerations of undrained strength of clays *(in Portuguese)*. *Solos e Rochas*, ABMS, ABGE, vol. 15 (2), 49–87.

Pinto, C. S. (1994). Lowland embankments *(in Portuguese)*. *Solos do Litoral do Estado de São Paulo*, ABMS, pp. 235–264.

Sandroni, S. S. and Bedeschi, M. V. R. (2008) Use of vertical drains in very soft clay at da Barra da Tijuca, Rio de Janeiro *(in Portuguese)*, *Proc. XIV Congresso Brasileiro de Mecânica dos Solos e Engenharia Geotécnica*, Búzios, on CD.

Sandroni, S. S. and Deotti, L. O. G. (2008). Instrumented test embankments on piles and geogrid platforms at the Panamerican Village, Rio de Janeiro. *1st Pan American Geosynthetics Conference & Exhibition*, Cancun, Mexico, 2–5 March 2008, on CD.

Sandroni, S. S. (1993). On the use of vane tests in embankment design *(in Portuguese)*. *Solos e Rochas*, 16 (3), 207–213.

Sandroni, S. S. (2006) The Brazilian practice of geotechnical design of highway embankments on very soft clays *(in Portuguese)*. *III Congresso Brasileiro de Geotecnia*, Curitiba, on CD.

Sandroni, S. S., Brugger, P. J. and Almeida, M. S. S. (1997). Geotechnical properties of Sergipe clay. In: Almeida, M. (ed.) *Recent Developments in Soil and Pavement Mechanics*, Rio de Janeiro, Brazil, pp. 271–277. Rotterdam: Balkema.

Sandroni, S. S., Lacerda, W. A. and Brant J. R. (2004). Stability control of embankments on soft clays by the volume method *(in Portuguese)* *Solos e Rochas*, 27 (1), pp. 25–35.

Saye, S. R. (2001). Assessment of soil disturbance by the installation of displacement sand drains and prefabricated vertical drains, *Geotechnical Special Publication*, ASCE, n.119.

Schnaid, F., Milittsky, J. and Nacci, D. (2001). Aeroporto Salgado Filho Airport – Civil Infra-struture and Geotechnical *(In Portuguese)*. 1. ed. Porto Alegre: Editora Sagras,. vol. 1, 260 p.

Vargas, M. (1973). Embankments in Santos lowlands *(in Portuguese)*. *Revista Politécnica*, Edição Especial, pp. 48–63.

Wegner, R., Candeias, M., Moormann, C., Jud, H. and Glockner, A. (2009). Soil improvement by stone columns for the ore storage yard at the Rio de Janeiro steel plant on soft, alluvial deposits. 17th *ICSMGE*, Alexandria, CD-ROM.

Zaeyen, V. D. B., Almeida, M. S. S., Marques, M. E. S. and Fuji, J. (2003). The behaviour of the embankment of Sarapuí Sewage Treatment Facility *(in Portuguese)*. *Solos e Rochas*, vol. 26 (3), pp. 261–271.

Soft soils improved by prefabricated vertical drains: performance and prediction

Indraratna, B. and Rujikiatkamjorn, C.
School of Civil, Mining and Environmental Engineering, Faculty of Engineering, University of Wollongong, Wollongong City, Australia

Wijeyakulasuriya, V.
Queensland Department of Main Roads, Brisbane, Australia

McIntosh, G.
Douglas Partners Pty Ltd, Unanderra, NSW Australia

Kelly, R.
Coffey Geotechnics, Sydney, Australia

KEYWORDS Analytical model, Cyclic loading, Numerical model, Soft soils, Vacuum preloading, Vertical drains

ABSTRACT The use of prefabricated vertical drains with preloading is now common practice and is proving to be one of the most effective ground improvement techniques known. The factors affecting its performance, such as the smear zone, the drain influence zone, and drain unsaturation, are discussed in this paper. In order to evaluate these effects, a large scale consolidation test was conducted and it was found that the proposed Cavity Expansion Theory could be used to predict the characteristics of the smear zone based on the soil properties available. Moreover, the procedure for converting an equivalent 2-D plane strain multi-drain analysis that considers the smear zone and vacuum pressure are also described. The conversion procedure was incorporated into finite element codes using a modified Cam-clay theory. Numerical analysis was conducted to predict excess pore pressure and lateral and vertical displacement. Three case histories are analysed and discussed, including the sites of Muar clay (Malaysia), the Second Bangkok International Airport (Thailand), and the Sandgate railway line (Australia). The predictions were then compared with the available field data, and they include settlement, excess pore pressure, and lateral displacement. The findings verified that smear and well resistance can significantly affect soil consolidation, which means that these aspects must be simulated appropriately to reliably predict consolidation using a selected numerical approach. Further findings verified that smear, drain unsaturation, and vacuum distribution can significantly influence consolidation, so they must be modeled appropriately in any numerical analysis to obtain reliable predictions.

1   INTRODUCTION

Preloading of soft clay with vertical drains is one of the most popular methods used to increase the shear strength of soft soil and control its post-construction settlement. Since the permeability of soils is very low, consolidation time to the achieved desired settlement or shear strength may take too long (Holtz,

1987; Indraratna et al., 1994). Using prefabricated vertical drains (PVDs), means that the drainage path is shortened from the thickness of the soil layer to the radius of the drain influence zone, which accelerates consolidation (Hansbo, 1981). This system has been used to improve the properties of foundation soil for railway embankments, airports, and highways (Li and Rowe, 2002).

Over the past three decades the performance of various types of vertical drains, including sand drains, sand compaction piles, prefabricated vertical drains (geosynthetic) and gravel piles, have been studied. Kjellman (1948) introduced prefabricated band shaped drains and cardboard wick drains for ground improvement. Typically, prefabricated band drains consist of a plastic core with a longitudinal channel surrounded by a filter jacket to prevent clogging. Most vertical drains are approximately 100 mm wide and 4 mm thick.

To study consolidation due to PVDs, unit cell analysis with a single drain surrounded by a soil cylinder has usually been proposed (e.g. Barron, 1948; Yoshikuni and Nakanodo, 1974). PVDs under an embankment not only accelerate consolidation, but they also influence the pattern of subsoil deformation. At the centre line of an embankment where lateral displacement is negligible, unit cell solutions are sufficient but elsewhere, especially towards the embankment toe, any prediction from a single drain analysis is not accurate enough because of lateral deformation and heave (Indraratna, et al., 1997).

Fig. 1 shows the vertical cross section of an embankment stabilised by a vertical drain system, with the instruments required to monitor the soil foundation. Before PVDs are installed superficial soil must be removed to ease the installation of the horizontal drainage, the site must be graded, and a sand platform compacted. The sand blanket drains water from the PVDs and supports the vertical drain installation rigs.

Fig. 2 illustrates a typical embankment subjected to vacuum preloading (membrane system). Where a PVD system is used with vacuum preloading, horizontal drains must be installed after a sand blanket has been put in place

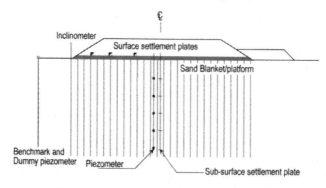

FIG. 1 *Vertical drain system with preloading*

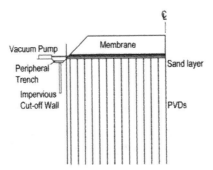

FIG. 2 *Vacuum preloading system*

(Cognon et al., 1994). The horizontal drains are connected to a peripheral bentonite slurry trench, which is then sealed with an impermeable membrane and cut-off walls to prevent possible vacuum loss at the embankment edges. The vacuum pumps are connected to the discharge module extending from the trenches. The vacuum generated by the pump increases the hydraulic gradient towards the drain which accelerates the dissipation of excess pore water pressure.

2   VERTICAL DRAIN CHARACTERISTICS

2.1   Equivalent drain diameter and drain influence zone

As shown in Fig. 3, PVDs with a rectangular cross section are usually installed in a triangular or square pattern. Their shapes are not the same as the circular cross section considered in the unit cell theory, so a PVD with a polygon influence zone must be transformed into a cylindrical drain with a circular influence zone (Fig. 4). The approximate equations proposed for the equivalent drain diameter are based on various hypotheses, hence the different results. The formulations for an equivalent cylindrical drain conversion available from previous studies are highlighted below:

$$d_w = 2(w + t)/\pi \qquad \text{(Hansbo, 1979)} \qquad (1)$$

$$d_w = (w + t)/2 \qquad \text{(Atkinson and Eldred, 1981)} \qquad (2)$$

$$d_w = 0.5w + 0.7t \qquad \text{(Long and Covo, 1994)} \qquad (3)$$

where d_w = equivalent PVDs diameter and w and t = width and thickness of the PVD, respectively.

2.2   Smear zone

The smear zone is the disturbance that occurs when a vertical drain is installed using a replacement technique. Because the surrounding soil is compressed during installation, there is a substantial reduction in permeability around the drain, which retards the rate of consolidation. In this section the

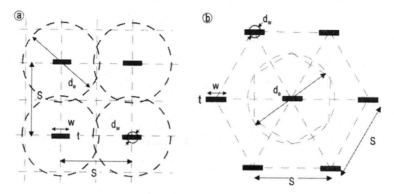

FIG. 3 *Drain installation pattern (a) square pattern; (b) triangular pattern*

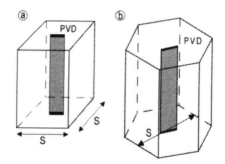

FIG. 4 *Vertical drain and its dewatered soil zone (a) unit cell with square grid installation and (b) unit cell with triangular grid installation*

Elliptical Cavity Expansion Theory was used to estimate the extent of the smear zone (Ghandeharioon et al. 2009; Sathananthan et al. 2008). This prediction was then compared with laboratory results based on permeability and variations in the water content. The detailed theoretical developments are explained elsewhere by Cao et al. (2001) and Ghandeharioon et al. (2009), so only a brief summary is given here. The yielding criterion for soil obeying the MCC model is:

$$\eta = M \sqrt{(p_c'/p') - 1} \tag{4}$$

where p_c' = the stress representing the reference size of yield locus, p'= mean effective stress, M = slope of the critical state line and η = stress ratio. Stress ratio at any point can be determined as follows:

$$\ln\left(1 - \frac{\left(a^2 - a_0^2\right)}{r^2}\right) = -\frac{2(1+v)}{3\sqrt{3}(1-2v)}\frac{\kappa}{v}\eta - 2\sqrt{3}\frac{\kappa\Lambda}{vM}f(M,\eta,OCR) \tag{5}$$

$$f(M, \eta, OCR) = \frac{1}{2} \ln \left[\frac{(M + \eta)\left(1 - \sqrt{OCR - 1}\right)}{(M - \eta)\left(1 + \sqrt{OCR - 1}\right)} \right]$$

$$- \tan^{-1}\left(\frac{\eta}{M}\right) + \tan^{-1}\left(\sqrt{OCR - 1}\right) \quad (6)$$

In the late expression, a = radius of the cavity, a_0 = initial radius of the cavity, ν = Poisson's ratio, κ = slope of the over consolidation line, υ = specific volume, OCR = over consolidation ratio and λ is the slope of the normal consolidation line).

Fig. 7 shows the variation of the permeability ratio (k_h/k_v), obtained from large scale laboratory consolidation and predicted plastic shear strain along the radius. Here the radius of the smear zone was approximately 2.5 times the radius of the mandrel, which agreed with the prediction using the cavity expansion theory.

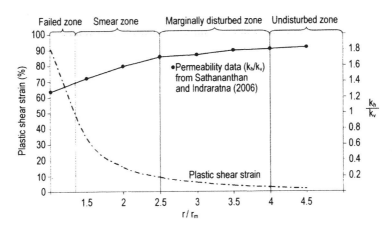

FIG. 5 *Variations in the ratio of the horizontal coefficient of permeability to the vertical coefficient of permeability and the plastic shear strain in radial direction (adopted from Ghandeharioon et al. 2009)*

2.3 Drain unsaturation

Due to an air gap from withdrawing the mandrel, and dry PVDs, unsaturated soil adjacent to the drain can occur. The apparent delay in pore pressure dissipation and consolidation can be observed during the initial stage of loading (Indraratna et al., 2004). Fig. 6 shows how the top of the drain takes longer to become saturated than the bottom. Fig. 6 illustrates the change in degree of saturation with the depth of the drain. Even for a drain as short as 1 m, the time lag for complete drain saturation can be significant.

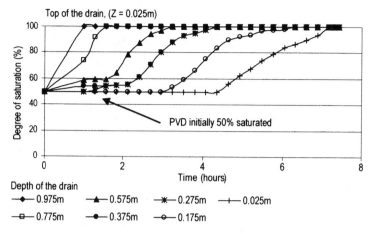

FIG. 6 *Degree of drain saturation with time (after Indraratna et al. 2004)*

3   EQUIVALENT PLANE STRAIN FOR MULTI-DRAIN ANALYSIS

In order to reduce the calculation time, most available finite element analyses on embankments stabilised by PVDs are based on a plane strain condition. To obtain a realistic 2-D finite element analysis for vertical drains, the *equivalence* between a plane strain condition and an *in situ* axisymmetric analysis needs to be established. Indraratna and Redana (2000); Indraratna et al. (2005) converted the unit cell of a vertical drain shown in Fig. 7 into an equivalent parallel drain well by determining the coefficient of permeability of the soil.

By assuming that the diameter of the zone of influence and the width of the unit cell in a plane strain to be the same, Indraratna and Redana (2000) presented a relationship between k_{hp} and k'_{hp}, as follows:

$$k_{hp} = \frac{k_h \left[\alpha + (\beta)\frac{k_{hp}}{k'_{hp}} + (\theta)\left(2lz - z^2\right)\right]}{\left[\ln\left(\frac{n}{s}\right) + \left(\frac{k_h}{k'_h}\right)\ln(s) - 0.75 + \pi\left(2lz - z^2\right)\frac{k_h}{q_w}\right]} \tag{7}$$

In Eq. 7, if well resistance is neglected, the smear effect can be determined by the ratio of the smear zone permeability to the undisturbed permeability, as follows:

$$\frac{k'_{hp}}{k_{hp}} = \frac{\beta}{\frac{k_{hp}}{k_h}\left[\ln\left(\frac{n}{s}\right) + \left(\frac{k_h}{k'_h}\right)\ln(s) - 0.75\right] - \alpha} \tag{8}$$

$$\alpha = \frac{2}{3} - \frac{2b_s}{B}\left(1 - \frac{b_s}{B} + \frac{b_s^2}{3B^2}\right) \tag{8a}$$

$$\beta = \frac{1}{B^2}\left(b_s - b_w\right)^2 + \frac{b_s}{3B^3}\left(3b_w^2 - b_s^2\right) \tag{8b}$$

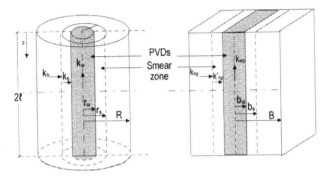

FIG. 7 *Conversion of an axisymmetric unit cell into plane strain condition (after Indraratna and Redana, 2000)*

$$\theta = \frac{2k_{hp}^2}{k'_{hp} q_z B}\left(1 - \frac{b_w}{B}\right) \tag{8c}$$

where k_{hp} and k'_{hp} are the undisturbed horizontal and the corresponding smear zone equivalent permeability, respectively.

The simplified ratio of plane strain to axisymmetric permeability by Hird et al. (1992) is readily obtained when the effect of smear and well resistance are ignored in the above expression, as follows:

$$\frac{k_{hp}}{k_h} = \frac{0.67}{[\ln(n) - 0.75]} \tag{9}$$

The well resistance is derived independently and yields an equivalent plane strain discharge capacity of drains, which can be determined from the following equation:

$$q_z = \frac{2}{\pi B}q_w \tag{10}$$

With vacuum preloading, the equivalent vacuum pressures in plane strain and axisymmetric are the same.

4   APPLICATION TO CASE HISTORIES

4.1 Muar clay embankment

One of the test embankments on Muar plain was constructed to failure. The failure was due to a "quasi slip circle" type of rotational failure at a critical embankment height at 5.5 m, with a tension crack propagating through the crust and the fill layer (Fig. 8). Indraratna et al. (1992) analysed the performance of the embankment using a Plane strain finite element analysis employing two distinct constitutive soil models, namely, the Modified Cam-clay theory using the finite element program CRISP (Woods, 1992) and the hyperbolic stress-strain behaviour using the finite element code ISBILD (Ozawa and Duncan,

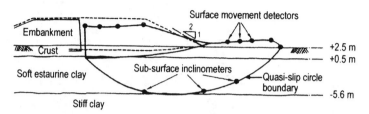

FIG. 8 *Failure mode of embankment and foundation (modified after Brand and Premchitt, 1989)*

1973). Two modes of analysis were used, undrained and coupled consolidation. Undrained analysis was used when the loading rate was much *faster* than the dissipation rate of excess pore pressure. This will cause excess pore pressure to build up during loading but will not alter the volume. While excess pore pressure is generated simultaneously with drainage, a positive or negative change in volume is allowed for coupled consolidation analysis.

The essential soil parameters used for the Modified Cam-clay model are summarised in Table 1 and a summary of soil parameters for undrained and drained analyses by ISBILD is tabulated in Table 2. Because properties of a topmost crust were not available, it was assumed that the soil properties were similar to the layer immediately below. The properties of the embankment surcharge ($E = 5,100\,\text{kPa}$, $\nu = 0.3$ and $\gamma = 20.5\,\text{kN/m}^3$), and related shear strength parameters ($c' = 19\,\text{kPa}$ and $\phi' = 26^0$), were obtained from drained triaxial tests.

TABLE 1 SOIL PARAMETERS USED IN THE MODIFIED CAM-CLAY MODEL (CRISP) (SOURCE: INDRARATNA ET AL.,1992)

Depth (m)	κ	λ	M	e_{cs}	$K_w \times 10^4$ (cm^2/s)	γ (kN/m^3)	$k_h \times 10^{-9}$ (m/s)	$k_v \times 10^{-9}$ (m/s)
0 – 2.0	0.05	0.13	1.19	3.07	4.4	16.5	1.5	0.8
2.0 – 8.5	0.05	0.13	1.19	3.07	1.1	15.5	1.5	0.8
8.5 – 18	0.08	0.11	1.07	1.61	22.7	15.5	1.1	0.6
18 – 22	0.10	0.10	1.04	1.55	26.6	16.1	1.1	0.6

TABLE 2 SOIL PARAMETERS FOR HYPERBOLIC STRESS STRAIN MODEL ISBILD (SOURCE: INDRARATNA ET AL.,1992)

Depth (m)	K	c_u (kPa)	K_{ur}	c' (kPa)	ϕ' (degree)	γ (kN/m^3)
0 – 2.5	350	15.4	438	8	6.5	16.5
2.5 – 8.5	280	13.4	350	22	13.5	15.5
8.5 – 18.5	354	19.5	443	16	17.0	15.5
18.5 – 22.5	401	25.9	502	14	21.5	16.0

Note: K and K_{ur} are the modulus number and unloading-reloading modulus number used to evaluate the compression and recompression of the soil, respectively.

The finite element discretisation is shown by Fig. 9. The embankment was constructed at a rate of 0.4 m/week. Instruments such as inclinometers, piezometers, and settlement plates were installed at this site (Fig. 10).

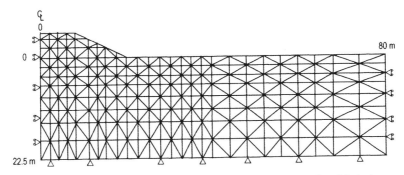

FIG. 9 *Finite element discretisation of embankment and subsoils (modified after Indraratna et al., 1992)*

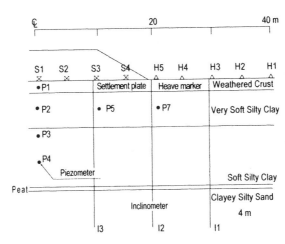

FIG. 10 *Cross section of Muar test embankment indicating key instruments (modified after Ratnayake, 1991)*

The yielding zones and potential failure surface observed were based on the yielded zone boundaries and maximum displacement vectors obtained from CRISP. Figs. 11 and 12 show the shear band predicted, based on the maximum incremental displacement and the boundaries of yielded zone approaching the critical state, respectively. The yielded zone was near the very bottom of the soft clay layer but it eventually spread to the centre line of the embankment, which verified that the actual failure surface was within the predicted shear band.

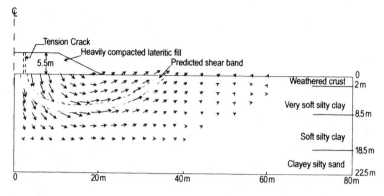

FIG. 11 *Maximum incremental development of failure (modified after Indraratna et al., 1992)*

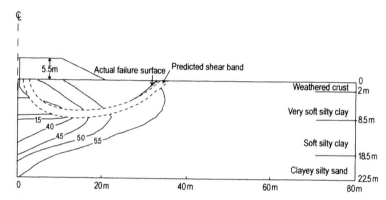

FIG. 12 *Boundary zones approaching critical state with increasing fill thickness (CRISP) (modified after Indraratna et al., 1992)*

4.2 Second Bangkok International Airport

The Second Bangkok International Airport or Suvarnabhumi Airport is about 30 km from the city of Bangkok, Thailand. Because the ground water was almost at the surface, the soil suffered from a very high moisture content, high compressibility and very low shear strength. The compression index ($C_c/(1 + e_0)$)varied between 0.2-0.3. The soft estuarine clays in this area often pose problems that require ground improvement techniques before any permanent structures can be constructed.

As reported by AIT (1995), the profile of the subsoil showed a 1 m thick, heavily over-consolidated crust overlying very soft estuarine clay, which was approximately 10 m below the bottom of a layer of crust. Approximately 10 to 21 m beneath this crust there was a layer of stiff clay. The ground water level varied from 0.5 to 1.5 m below the surface. The parameters of these layers of subsoil, based on laboratory testing, are given in Table 3.

TABLE 3 SELECTED SOIL PARAMETERS IN FEM ANALYSIS (INDRARATNA ET AL., 2005)

Depth (m)	λ	κ	ν	k_v 10^{-9} m/s	k_h 10^{-9} m/s	k_s 10^{-9} m/s	k_{hp} 10^{-9} m/s	k_{sp} 10^{-9} m/s
0.0 – 2.0	0.3	0.03	0.30	15.1	30.1	89.8	6.8	3.45
2.0 – 8.5	0.7	0.08	0.30	6.4	12.7	38.0	2.9	1.46
8.5 – 10.5	0.5	0.05	0.25	3.0	6.0	18.0	1.4	0.69
10.5 – 13	0.3	0.03	0.25	1.3	2.6	7.6	0.6	0.30
13.0 – 15	1.2	0.10	0.25	0.3	0.6	1.8	0.1	0.07

Two embankments stabilised by vacuum combined with surcharge loading (TV2) and surcharge loading alone (TS1) are described in this section.
The performances of embankments TV2 and TS1 were reported by Indraratna
and Redana (2000), and Indraratna et al. (2005), respectively. The vertical
cross section of Embankment TS1 is shown in Fig. 13. TS1 was constructed
in multi-stages, with 12 m long PVDs @ 1.5 m in a square pattern. The embankment was 4.2 m high with a 3H:1V side slope. Embankment TV2 was stabilised
with vacuum combined surcharge and 12 m long PVDs. A membrane system
was also used on this site.

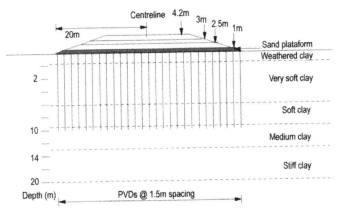

FIG. 13 *Cross section at embankment TS1 (after Indraratna and Redana, 2000)*

Both embankments were analysed using the finite element software
ABAQUS. The equivalent plane strain model (Eqs. 7-10) and modified Cam-
clay theory were incorporated into this analysis. The comparisons of the degree of consolidation based on settlement from the FEM and field measurement at the centre line of the embankment are presented in Fig. 14. It can be
seen that the application of vacuum pressure reduced the time from 400 to 120
days to achieve the desired degree of consolidation. Fig. 15 shows the time dependent excess pore water pressure during consolidation. The vacuum loading generated negative excess pore pressure in TV2 whereas the surcharge fill

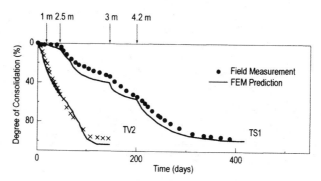

FIG. 14 *Degree of Consolidation at the centreline for embankments (after Indraratna and Redana, 2000, and Indraratna et al., 2005)*

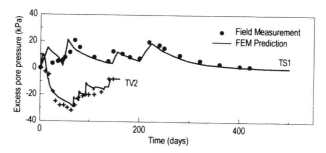

FIG. 15 *Excess pore pressure variation at 5.5 m depth (after Indraratna and Redana, 2000, and Indraratna et al., 2005)*

in embankment TS1 created a positive excess pore pressure. These predicted excess pore pressures agreed with the field measurements. The maximum negative excess pore pressure was approximately 40 kPa, probably caused by a puncture in the membrane and subsequent loss of air. The total applies stresses for both embankment were very similar and therefore yielded similar ultimate settlements (90 cm). The reduction in negative pore pressure at various times was caused by the vacuum being lowered. Despite these problems, the analysis using the proposed conversion procedure, including the smear effects, could generally predict the field data quite accurately.

4.3 Sandgate railway embankment

Under railway tracks where the load distribution from freight trains is typically kept below 7-8 m from the surface, relatively short PVDs may still dissipate cyclic pore pressures and curtail any lateral movement of the soft formation. It was expected that any excessive settlement of deep estuarine deposits during the initial stage of consolidation may compensate for continuous ballast packing. However, the settlement rate can still be controlled by optimising the spacing and pattern of drain installation. In this section, a case history where short PVDs were installed beneath a rail track built on soft formation is presented with the finite element analysis (Indraratna et al. 2009). The finite element analysis used by the Authors to design the track was a typical Class A prediction for a field observation because it was made before it was constructed.

To improve the conditions for rail traffic entering Sandgate, Kooragang Island, Australia, where major coal mining sites are located, two new railway lines were needed close to the existing track. An *in situ* and laboratory test was undertaken by GHD Longmac (Chan, 2005) to obtain the essential soil parameters. This investigation included boreholes, piezocone tests, *in situ* vane shear tests, test pits, and laboratory tests that included testing the soil index properties, standard oedometer testing, and vane shear testing.

The existing embankment fill at this site overlies soft compressible soil from 4 to 30 m deep over a layer of shale bedrock. The properties of this soil, with depth, are shown in Fig. 16, where the groundwater level was at the surface. Short, 8 m long PVDs were used to dissipate excess pore pressure and curtail lateral displacement. There was no preloading surcharge embankment provided due to stringent time commitments. The short PVDs were only expected to consolidate a relatively shallow depth of soil beneath the track where it would be affected by the train load. This initial load was considered to be the only external surcharge. An equivalent static approach based on the dynamic impact factor was used to simulate the field conditions, in this instance a static load of 80 kPa and an impact factor of 1.3 in conjunction with a speed of 40 km/h and a 25 tonne axle load. The Soft Soil model and Mohr-Coulomb model incorporated into the finite element code PLAXIS, were used in this analysis (Brinkgreve, 2002). Fig. 17 illustrates a cross-section of the rail track formation.

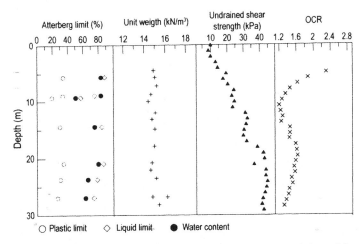

FIG. 16 *Soil properties at Sandgate Rail Grade Separation Project (adopted from Indraratna et al., 2009)*

In the field the 8 m long PVDs were spaced at 3 m intervals, based on the Authors' analysis and recommendations. Figs. 18 and 19 show a comparison between the predicted and measured settlement at the centre line of the rail track and lateral displacement after 180 days, respectively. The predicted settlement agreed with the field data for a Class A prediction, with the maxi-

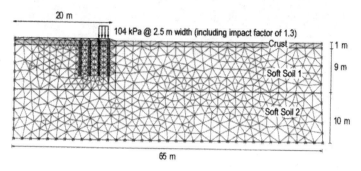

FIG. 17 *Vertical cross-section of rail track foundation (after Indraratna et al., 2009)*

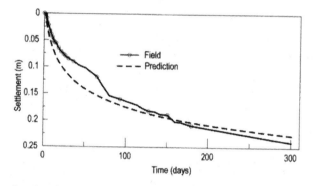

FIG. 18 *Predicted and measured at the centre line of rail tracks (after Indraratna et al., 2009)*

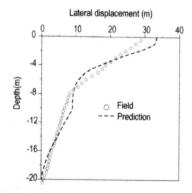

FIG. 19 *Measured and predicted lateral displacement profiles near the rail embankment toe at 180 days (after Indraratna et al., 2009)*

mum displacement being contained within the top layer of clay. The "Class A" prediction of lateral displacement agreed with what occurred in the field.

5   CONCLUSION

Various types of vertical drains have been used to accelerate the rate of primary consolidation. A comparison between embankments stabilised with a vacuum combined with a surcharge, and a surcharge alone, were analysed and discussed. Consolidation time with a vacuum applied was substantially reduced and lateral displacement curtailed, and if sufficient vacuum pressure is sustained, the thickness of the surcharge fill required may be reduced by several metres.

A plane strain finite element analysis with an appropriate conversion procedure is often enough to obtain an accurate prediction for large construction sites. An equivalent plane strain solution was used for selected case histories to demonstrate its ability to predict realistic behaviour. There is no doubt that a system of vacuum consolidation via PVDs is a useful and practical approach for accelerating radial consolidation because it eliminates the need for a large amount of good quality surcharge material, via air leak protection in the field. Accurate modelling of vacuum preloading requires both laboratory and field studies to quantify the nature of its distribution within a given formation and drainage system.

It was shown from the Sandgate case study that PVDs can decrease the buildup of excess pore water pressure during cyclic loading from passing trains. Moreover, during rest periods PVDs continue to simultaneously dissipate excess pore water pressure and strengthen the track. The predictions and field data confirmed that lateral displacement can be curtailed, which proved that PVDs can minimize the risk of undrained failure due to excess pore pressure generated by cyclic train loads.

ACKNOWLEDGEMENTS

The authors appreciate the support given by the Australian Rail Track Corporation (ARTC), and John Holland Pty Ltd. They wish to thank the CRC for Rail Innovation (Australia) for its continuous support. The embankment data provided by the Asian Institute of Technology are appreciated. A number of other current and past doctoral students, namely, Mr. Somalingam Balachandran, Ms. Pushpachandra Ratnayake, Dr. I Wayan Redana, Dr. Chamari Bamunawita, Dr. Iyathurai Sathananthan, Dr. Rohan Walker, and Mr. Ali Ghandeharioon are also contributed to the contents of this keynote paper. More elaborate details of the contents discussed in this paper can be found in previous publications of the first author and his research students in Geotechnique, ASCE, Canadian Geotechnical Journals, since mid 1990's and Dr. Rujikiatkamjorn PhD thesis for the work related to Bangkok case histories.

REFERENCES

AIT. (1995). The Full Scale Field Test of Prefabricated Vertical Drains for The Second Bangkok International Airport (SBIA). AIT, Bangkok, Thailand.

Atkinson, M. S. and Eldred, P. (1981). Consolidation of soil using vertical drains. *Geotechnique*, 31(1), 33-43.

Barron, R. A. (1948). Consolidation of fine-grained soils by drain wells. *Transactions ASCE*, Vol. 113, pp. 718-724.

Brand, E. W. and Premchitt, J. (1989). Moderator's report for the predicted performance of the Muar test embankment. *Proc. International Symposium on Trial Embankment on Malysian Marine Clays*, Kuala Lumpur, Malaysia, Vol. 2, pp. 1/32-1/49.

Brinkgreve, R. B. J. (2002). PLAXIS (Version 8) User's Manual. Delft University of Technology and PLAXIS B.V., Netherlands.

Cao, L. F., Teh, C. I., and Chang, M. F. (2001). Undrained Cavity Expansion in Modified Cam Clay I: Theoretical Analysis. *Geotechnique*, Vol. 51(4); pp. 323-334.

Chan, K. (2005). Geotechnical information report for the Sandgate Rail Grade Separation, Hunter Valley Region, Australia.

Cognon, J. M., Juran, I. and Thevanayagam, S. (1994). Vacuum consolidation technology-principles and field experience. *Proc. Conference on Foundations and Embankments Deformations*, College Station, Texas, Vol. 2, pp.1237-1248.

Ghandeharioon, A., Indraratna, B., and Rujikiatkamjorn, C. (2009). Analysis of soil disturbance associated with mandrel-driven prefabricated vertical drains using an elliptical cavity expansion theory. *International Journal of Geomechanics*, ASCE. (Accepted, August 2009).

Hansbo, S. (1981). Consolidation of fine-grained soils by prefabricated drains. *In Proceedings of 10th International Conference on Soil Mechanics and Foundation Engineering*, Stockholm, Balkema, Rotterdam, 3, pp. 677-682.

Hird, C. C., Pyrah, I. C., and Russell, D. (1992). Finite element modelling of vertical drains beneath embankments on soft ground. *Geotechnique*, Vol. 42(3), pp. 499-511.

Holtz, R. D. (1987). Preloading with prefabricated vertical strip drains, *Geotextiles and Geomembranes*, Vol. 6 (1–3), pp. 109–131.

Indraratna, B., and Redana, I. W. (2000). Numerical modeling of vertical drains with smear and well resistance installed in soft clay. *Canadian Geotechnical Journal*, Vol. 37, pp. 132-145.

Indraratna, B., Bamunawita, C., and Khabbaz, H. (2004). Numerical modeling of vacuum preloading and field applications. *Canadian Geotechnical Journal*, Vol. 41, pp. 1098-1110.

Indraratna, B., Balasubramaniam, A. S. and Balachandran, S. (1992). Performance of test embankment constructed to failure on soft marine clay. *Journal of Geotechnical Engineering*, ASCE, Vol. 118(1), pp. 12-33.

Indraratna, B., Balasubramaniam, A. S., and Ratnayake, P. (1994). Performance of embankment stabilized with vertical drains on soft clay. *J. Geotech. Eng.*, ASCE, Vol. 120(2), pp. 257-273.

Indraratna, B., Balasubramaniam, A. S. and Sivaneswaran, N. (1997). Analysis of settlement and lateral deformation of soft clay foundation beneath two full-scale embankments. *International Journal for Numerical and Analytical Methods in Geomechanics*, Vol. 21, pp. 599-618.

Indraratna, B., Rujikiatkamjorn C., and Sathananthan, I. (2005). Analytical and numerical solutions for a single vertical drain including the effects of vacuum preloading. *Canadian Geotechnical Journal*, Vol.42, pp. 994-1014.

Indraratna, B., Rujikiatkamjorn, C. Adams, M., and Ewers, B. (2009). Class A prediction of the behaviour of soft estuarine soil foundation stabilised by short vertical drains beneath a rail track. *International Journal of Geotechnical and Geo-environmental Engineering*, ASCE (Accepted October 2009).

Kjellman, W. (1948). Accelerating consolidation of fine grain soils by means of cardboard wicks. *Proc. 2nd ICSMFE*, Vol. 2, pp. 302-305.

Li, A. L. and Rowe, R. K. (2002). Combined effect of reinforcement and pre-fabricated vertical drains on embankment performance. *Canadian Geotechnical Journal*, Vol. 38, pp. 1266-1282.

Long, R. and Covo, A. (1994). Equivalent diameter of vertical drains with an oblong cross section. *Journal of Geotechnical Engineering*, Vol.20(9), 1625-1630.

Ozawa, Y. and Duncan, J. M. (1973). ISBILD: A computer program for static analysis of static stresses and movement in embankment. University of California, Berkeley, California.

Ratnayake, A. M. P. (1991). Performance of test embankments with and without vertical drains at Muar flats site, Malaysia. *Master Thesis*, GT90-6, Asian Institute of Technology, Bangkok.

Sathananthan, I. and Indraratna, B. (2006). Laboratory Evaluation of Smear Zone and Correlation between Permeability and Moisture Content, *J. of Geotechnical & Geoenvironmental Engineering*, ASCE, Vol. 132(7), pp. 942-945.

Sathananthan, I., Indraratna, B., and Rujikiatkamjorn C., (2008). The evaluation of smear zone extent surrounding mandrel driven vertical drains using the cavity expansion theory. *International Journal of Geomechanics*, ASCE. Vol. 8(6), 355-365.

Woods, R. (1992). SAGE CRISP technical reference manual. The CRISP Consortium Ltd. UK.

Yoshikuni, H. and Nakanodo, H. (1974). Consolidation of fine-grained soils by drain wells with finite permeability. *Japanese Society Soil Mechanics and Foundation Engineering*, Vol. 14(2), pp. 35-46.

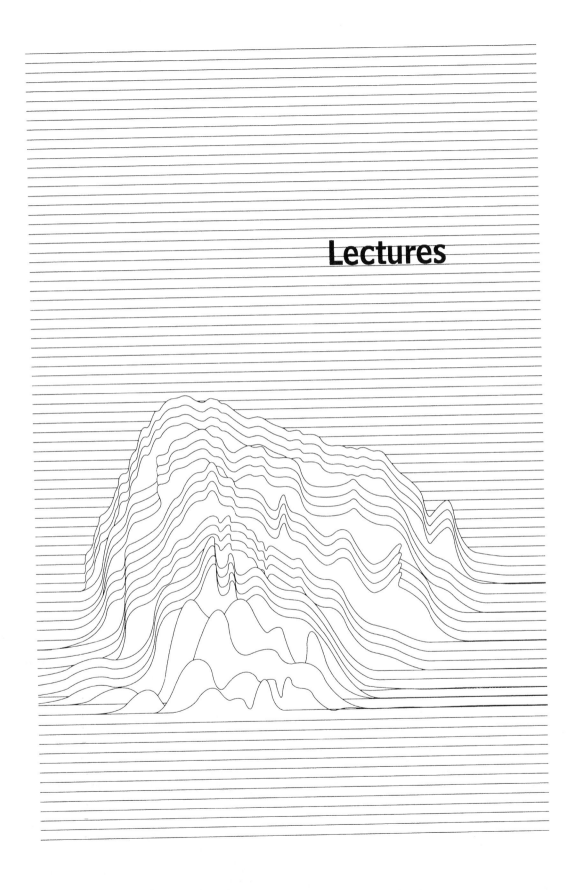

Lectures

Analysis and Control of the Stability of Embankments on Soft Soil: Juturnaíba and other Experiences in Brazil

Coutinho, R. Q. & Bello, M. I. M. C

Department of Civil Engineering, Federal University of Pernambuco, Brazil

KEYWORDS monitoring, stability of embankments, soft soils

ABSTRACT In the design and evaluation of the behavior of embankments on soft clay foundation, geotechnical parameters and the instrumentation of pore pressure and displacements (vertical and horizontal) are required in order to have an efficient construction. A summary of results of stability analysis and stability control, performed by the Geotechnical Group (GEGEP) of the Federal University of Pernambuco, Brazil is presented in this paper. Three case studies were conducted in Brazil: the Juturnaíba trial embankments in Rio de Janeiro, the access embankments of the Jitituba River Bridge in Alagoas and the failure of an embankment on the BR-101-PE highway, located in Recife, Pernambuco. The stability analysis and stability control studies presented satisfactory results when compared to those in the literature given the geotechnical characteristics of each case.

1 INTRODUCTION

The construction of embankments on soft clay raises an important geotechnical problem which has been studied by various authors, thus accumulating experiences for a better understanding of soft soils subjected to load increases (e.g. Bjerrum, 1973; Tavenas and Leroueil, 1980; Ladd, 1991; Leroueil and Rowe, 2000). In Brazil, among important research studies are those by Ortigão (1980), Coutinho (1986) and Magnani de Oliveira (2006). In general, the design of embankments on soft soils should meet the basic requirements of stability against rupture and displacement (vertical and horizontal), during and after construction, compatible with its objective. The use of instrumentation is a tool for monitoring and evaluating (including stability control) the construction of embankments by means of which pore-pressures and vertical and horizontal displacements are measured.

This paper presents a summary of the stability analysis and control results of the Juturnaíba trial embankment and three other studies conducted by GEGEP, the Geotechnical Group of Federal University of Pernambuco.

2 CASE STUDIES

The Juturnaíba Dam Project, an earth-fill structure located in the north of Rio de Janeiro State, Brazil, was built in 1981-1983. The foundation consisted of an organic clay deposit about 8 m thick, with SPT values ranging from 0/111 to 1/33, typically 0/50, along its full depth, underlain by sand sediments with SPT values about 10/30 to a depth of 14 m. Visual classification and laboratory tests permitted a division of the clayey deposit into six layers, with varying organic and water content, ranging from light-grey silt clay to brown clayey peat. Geotechnical studies were quite comprehensive and included laboratory and fields investigations as well as the construction of a trial embankment taken to failure, which was very well instrumented as indicated in Fig. 1 (Coutinho, 1986; Coutinho and Lacerda, 1987, 1989). The main purposes of these studies were to provide indications on the undrained strength and compressibility in the clay foundation and on methods to control stability during the construction.

The second case presents the study on the access embankments of the Jitituba River Bridge, located on the Alagoas- 413 Highway, where this bridge was built before the access embankments. Due to there being a (12 m thick) soft soil layer and to the construction sequence of the bridge on the embankments of the Jitituba River, there was a need to analyze the vertical and horizontal displacements and the consequent efforts to transverse the piles through the bridge (Fig. 2). The behavior of the access embankments was analyzed in relation to the measurements of pore-pressures, and vertical and horizontal displacements, by applying models proposed in the literature and by comparison with other case studies of embankments on soft soils. The solution adopted consisted of constructing embankments in stages, allied to the use of prefabricated vertical drains and geotechnical instrumentation

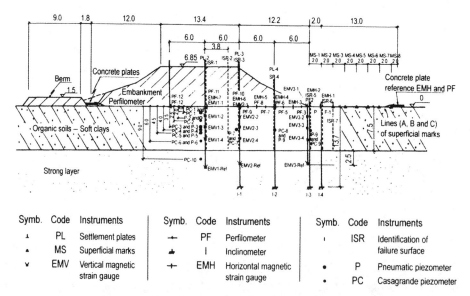

FIG. 1 *Instrumentation of the Juturnaíba trial embankment (Coutinho, 1986)*

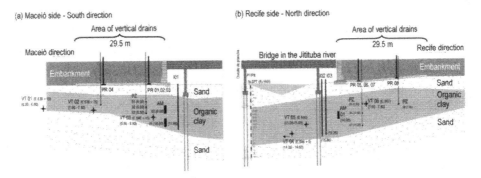

FIG. 2 *Longitudinal section from the basic project for the access embankments of the bridge over Jitituba River, including the geotechnical profile and the location of the field investigations (Cavalcante, 2001; Cavalcante et al., 2004)*

(Casagrande piezometers, settlement plates and inclinometers) to control and monitor the project performance (Cavalcante, 2001; Cavalcante et al., 2003; 2004).

The third case is a rupture of an embankment on soft clays that occurred in an area beside the federal BR-101-Pernambuco highway (Bello, 2004; Coutinho and Bello, 2005; Bello et al., 2006). Fig. 3 shows the position of the sheds and the location of the geotechnical field investigations. The geotechnical profile presents a thick layer of soft soil which is about 13 m thick. The

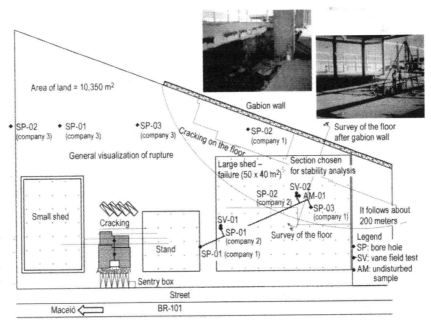

FIG. 3 *Situation and localization of SPT soundings, vane field tests and undisturbed sampling – Embankment beside the BR-101-PE highway (Bello, 2004)*

embankment was constructed without a geotechnical investigation program, monitoring plan or technological control. After the failure, in order to understand the process, *in situ* and laboratory tests was performed so as to permit total stress stability analysis / back-analysis. The circular and non-circular surfaces, the cracking in embankment, and the three-dimensional effect were examined. The Data Base of Recife Soft Clays was used to complement the technical information necessary. The research studies on the second and third cases were made possible by there being a partnership with Gusmão Engineer Associated.

3 STABILITY ANALYSIS

Ladd (1991) defined three types of stability analysis for embankments on soft soils: (a) total stress analysis (TSA); (b) undrained strength analysis (USA), and (c) effective stress analysis (ESA). Total stress analysis is often used in single-stage construction analysis and is usually based on the undrained strength profile prior to construction. In undrained strength analysis, the *in situ* undrained shear strength is computed as a function of the pre-shear effective stress. This type of analysis is often used in evaluating the stability of embankments constructed in stages.

Evaluation of mobilized shear strength Su in an embankment constructed in one stage can be obtained by several approaches: (a) the field vane test approach, with the Bjerrum (1973) correction factor, μ; (b) pre-consolidation pressure $Su/\sigma_p' = 0.19$ (plasticity index PI = 10%) to about 0.28 (PI = 80%). The upper values often correspond to organic clays; (c) recompression and SHANSEP approaches; (d) the direct simple shear test; (e) the unconfined and unconsolidated undrained compression test and (f) Cone penetration tests and Marchetti Dilatometer tests. To gain confidence in the results of stability analyses, it is recommended that at least two of these approaches be considered in practical applications. In the case of an embankment constructed in several stages, the choice of strength can be obtained using several approaches: (a) field vane test approach, without the Bjerrum (1973) correction; (b) CPT tip resistance approach; (c) vertical effective stress approach ($Su_v/\sigma_{vc}' = 0.25$); (d) SHANSEP approach (Leroueil and Rowe, 2000). In the effective stress analysis approach, mobilized strength parameters are close to those of normally consolidated clay.

3.1 Results – Total stress analysis

Coutinho (1986) performed a total stress stability analysis on the Juturnaíba trial embankment to obtain the minimal factor of safety, FS_{min}. The principal analysis was performed using the Modified Bishop method, which takes account of the circular surface. The Modified Janbu method was utilized in a complementary analysis (back-analysis for the failure surface), which took into account that there was no circular surface. In the study, 7 hypotheses were established considering the *Su* profile obtained from the *in situ* vane test (average ± standard deviation) and the cracking of the embankment (the embankment strength is not mobilized). The analysis was performed for heights of 6.85 m at which the failure of the foundation occurred (intense "instantaneous" displacements etc.) and also for the height of 8.85 m so as to evaluate the behavior of the embankment in the phase after the failure, which occurred while the embankment continued to be constructed.

Fig. 4 shows *Su* values obtained in the field vane test and in triaxial UU and CIU ($\sigma'_c \cong \sigma'_{oct}$ *in situ*) tests as well as the mean values and the range of field vane shear tests results with the equations for each of the six soil layers. The following points emerged from analyzing a comparison of the data: 1) Individual values and linear regression of the *Su* field vane test (Fig. 4a) were basically distinct for each layer, and showed in layers I and II a certain variation. In the tests with "remolded" soil, values of *Su* were low, with a sensitivity near or equal to 10, with great dispersion and little variation among the layers; 2) Su values from UU and CIU tests are very similar and fall close to mean vane shear strength results (Fig. 4b). Results from CIU tests present smaller dispersion than the UU values; 3) *Su* results in layer III from the triaxial tests are practically constant with the depth, which accords well with results of maximum past preconsolidation pressure (Coutinho and Lacerda, 1987). The Mesri (1975) proposal for the "mobilized" Su = $0.22\sigma'_p$ is also presented in Fig. 4b, showing smaller results, as expected, than those obtained directly from the triaxial and vane tests.

The stability analysis for the 6.85 m high embankment (failure condition) considering the hypothesis of average *Su* from the field vane test performed

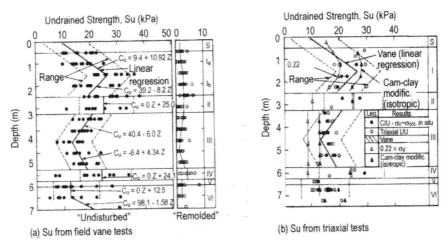

(a) Su from field vane tests

(b) Su from triaxial tests

FIG. 4 *Undrained strength values vs. depth: (a) field vane test; (b) triaxial tests (Coutinho and Lacerda, 1989)*

before construction, without the Bjerrum correction, obtained values of 1.069, 1.001 and 0.960 for FS_{min}, depending on the consideration of cracking in embankment, hypotheses 1, 4 and 5 respectively (Table 1).

With the consideration of the Su vane range (mean values ± standard deviation) and the strength of embankment (0% cracking), the results of the factors of safety obtained were 1.264 and 0.888, respectively. The use of the Bjerrum correction showed very low results ($FS \ll 1.0$), considering the average high plasticity of the soft deposit. The back-analysis realized using the observed failure surface showed satisfactory results of FS_{min}, displaying values close to the preliminary analysis (5 to 9% higher).

Fig. 5 presents FS_{min} results for heights of embankment considering the hypothesis of average Su from the field vane test and the effect of cracking of the embankment in the results. An appreciable reduction in FS_{min} value can be observed with the continuity of the loading, particularly at embankment heights of over 5.65 m ($FS_{min} = 1.31$). The influence of cracks in the embankment on the FS_{min} values was in the order of 10%. The stability analysis for the 8.85 m high embankment (construction post failure) showed results of FS around 1.0, confirming the rupture condition. The different behavior of the Juturnaíba foundation, which did not need the Bjerrum correction) can be explained by the organic condition of the soil deposit and the strong drainage with significant deformation and increase of stress during construction. The analysis also shows that the Mesri (1975) proposal is not adequate to represent the mobilized strength.

To understand the failure process better, after the construction of the 8.85 m high embankment was completed, the embankment was excavated,

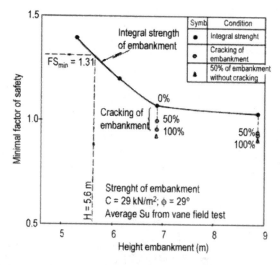

FIG. 5 *Summary of the total stability analysis results (Coutinho, 1986)*

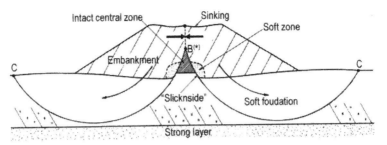

FIG. 6 *Shared failure of the Juturnaíba Trial embankment (Coutinho, 1986)*

observing its condition and the foundation during the work. Fig. 6 presents what was observed showing a shared failure, for both sides, with a slick side zone in the foundation and a loose zone in the embankment. Because of this type of shared failure, what was not observed was the traditional movement of an embankment volume in one horizontal direction in the failure in the 6.85 m high embankment.

Bello (2004) performed total stability analysis in an embankment on soft soil (third case of this paper). The study sets out to understand the failure and to confirm the necessity to correct Su from the vane field test on Recife soft clays. Sub layers with different type of soils and their respective Su values obtained from field vane tests corrected by the Bjerrum (1973) proposal were defined. The calculation of FS_{min} was made using the Modified Bishop, Janbu, Spencer and Morgenstern-Price methods. Table 1 shows a summary of FS_{min} results obtained from stability analysis and back-analysis for three hypotheses on the cracking of the embankment and the correct Su value. In the stability analysis, the FS_{min} values are in the range of 0.995 to 1.082 for the circular surface condition, depending on the strength of the embankment and consideration of Su. Hypothesis 3 (embankment 50% cracking and Su corrected) presented FS_{min} equal to 1.00 explaining the rupture (Fig. 7). The influence of cracks in the embankment on the FS_{min} values was in the order of 10 to 15%. In the back-analysis FS_{min} results indicated a range of values close to the preliminary stability analysis (around 15% higher). This difference may be due to the difficulty of defining the failure surface in the field.

In the Juturnaíba test embankment, the predicted critical circle was very similar to the failure surface observed *in situ*, and was only slightly dislocated to the left. Bello (2004) observed similar results for the predicted critical circle in the embankment on the studied soft soil.

The stability study was also performed using empirical methods (Load Capacity Equation; the Sliding Wedge Method; Chart of Pillot and Moreau; Chart of Pinto, using representative Su for each case. Table 2 shows results obtained from the Recife, Juturnaíba and Sarapuí deposits. The results were satisfactory, which stimulates the applicability of these methods for preliminary

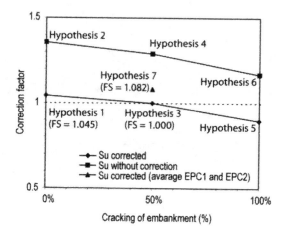

FIG. 7 *Comparison of results of stability analysis / back-analysis (circular surface) – Bishop Method (Bello, 2004)*

results, particularly the load capacity and sliding wedge methods. The three-dimensional effect (Azzouz et al., 1983) showed an insignificant increase in the order of 10% in relation to the bi-dimensional FS in the Juturnaíba deposit and in the order of 5% for the one in Recife for the critical factor of safety.

Fig. 8 presents the Bjerrum (1973) and Azzouz et al. (1983) proposal for the correction factor, μ, to be applied in *Su* from field vane test to obtain the undrained strength for design. It is also shown that the Brazilian results validated the Bjerrum proposal (in general the two proposals), presenting a general average result of μ is 0.65. The Juturnaíba trial embankment results lie outside the proposal as it has an organic soil foundation.

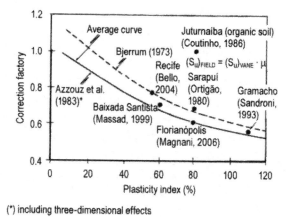

(*) including three-dimensional effects

FIG. 8 *Factors of correction from back-analysis of rupture embankments (Bello, 2004; Coutinho and Bello, 2005; Coutinho, 2006; Sandroni, 2006)*

TABLE 1 SUMMARY OF THE FSMIN CALCULATED

Case studied	Hypotheses	Cracking of Embankment (%) 0	50	100	Su Correction	Su Average	Su standard deviation	CIRCULAR SURFACE ANALYSIS BISHOP	SPENCER	NO CIRCULAR SURFACE BACK-ANALYSIS BISHOP	JANBU	SPENCER	MORGENSTERN-PRICE
(Case 2) Juturnaíba trial embankment H_{emb} = 6.85 m - Coutinho (1986)	1	X	—	—	—	X	—	1.069	—	1.165	1.205	—	—
	2	X	—	—	—	X	(-) X	0.888	—	—	—	—	—
	3	X	—	—	—	X	(+) X	1.274	—	—	—	—	—
	4	—	X	—	—	X	—	1.001	—	1.046	—	—	—
	5	—	—	X	—	X	—	0.960	—	1.007	—	—	—
	6	—	X	—	—	X	(+) X	1.200	—	—	—	—	—
	7	—	—	X	—	X	(+) X	1.195	—	—	—	—	—
(Case 4) Embankment beside the BR-101-PE highway - Bello (2004)	1	X	—	—	X	—	—	1.045	1.048	—	1.240	1.232	1.232
	3	—	X	—	X	—	—	1.000	0.995	—	1.128	1.141	1.141
	7	—	X	—	X	—	—	1.082	1.076	—	1.153	1.186	1.186

TABLE 2 SUMMARY OF SU (BACK-ANALYSIS) AND FS VALUE OBTAINED FROM EMPIRICAL METHOD (COUTINHO AND BELLO, 2007)

METHOD	Height of embankment in a rupture (m) (1)	(2)	(3)	Su (kPa) back-analysis (FS = 1) (1)	(2)	(3)	representative Su value (kPa) (1)	(2)	(3)	FS representative Su (kPa) value (1)	(2)	(3)
Load Capacity Equation (Terzaghi, 1943)	2.5	6.85	6.0	9.4	19.7	19.63	9.0	19.0	20.59	1.07	0.96	1.05
Sliding Wedges (NAVFAC, 1971).				9.7	—	19.08				*	—	1.06
Pillot and Moreau (1973)				9.5	17.3	20.52				*	1.06	1.10
Abacus of Pinto (1966)				—	—	18.00				—	—	1.14

* The author did not calculate FS for Su representative value. Number of authors: (1) Sarapuí (Ortigão, 1980); (2) Juturnaíba (Coutinho, 1986); (3) Recife (Bello, 2004)

Specific weight of embankment: (1) $\gamma_{embank.}$ = 18.4 kN/m³; (2) $\gamma_{embank.}$ = 15.8 kN/m³; (3) $\gamma_{embank.}$ = 18.0 kN/m³

3.2 Results – Effective stress analysis

Coutinho (1986) performed an effective stress stability analysis on the Juturnaíba trial embankment to obtain the minimal factor of safety, for the height of the embankment at which failure occurs ($H_{emb} = 6.85$ m). This analysis was performed basically with the Modified Bishop method, using preconsolidated and normally consolidated effective parameters of strength and pore pressure measured and estimated by the pneumatic piezometer.

The analysis considering normally consolidated effective stress parameters of strength presented satisfactory results particularly when cracking of the embankment was considered to simulate the failure. The value obtained for FS_{min} ranged from 0.95 to 1.23. For the case of 50% of cracking of embankment the FS_{min} was 1.05.

The predicted critical circle for the effective analysis was distinct for the failure surface observed *in situ*. The predicted circle presented a smaller extension in area and in maximum depth. The studies realized considering the observed failure surface showed values for the factor of safety greater than the corresponding ones obtained in the study of the FS_{min}. For the case of 50% of cracking of the embankment and the same effective strength, the FS value was 1.169.

The estimation of values of pore pressures, in points where there are no piezometers and the difficulties of measuring the pore pressure at the moment of failure, can cause a reduction in the accuracy of an effective stress analysis.

4 STABILITY CONTROL

The stability control of an embankment can be performed during construction using the measures of displacements, deformations or pore pressures.

4.1 Pore Pressure

The stability control through pore pressures can be performed using interstitial pressure measures in an effective stability analysis. In this method, it is necessary to obtain the effective strength parameters, adequate measurement of pore pressures, and time needed for analyzing the results.

The results of increases in pore pressures observed near the middle of the soft deposit of foundation, under the center of the embankment, may display a substantial increase in the values observed with the proximity of the failure. In this case the pore pressure parameter B ($\Delta u / \Delta \sigma_v$) presents values greater than 1.0. According to Tavenas and Leroueil (1980) and Leroueil and Rowe (2001) this condition would be a signal of local failure. In the Juturnaíba trial embankment, this result was a signal indicating the beginning of the failure.

4.2 Horizontal Displacements

Table 3 presents a summary of the stability control proposal for horizontal displacements that are presented and discussed in this study. In the analysis, two experimental embankments induced to rupture were considered: Juturnaíba and Sarapuí, and two embankments designed to be stable: the Juturnaíba Dam and the Jitituba River Bridge. All behavior has to be analyzed, not just the values of the measures. It is also recommended that, in a practical case, more than one stability control proposal be used.

4.2.1 Tendency of the Horizontal Displacements

Fig. 9 presents the evolution of the maximum horizontal (lateral) displacements with time considering the Jitituba River embankment construction and the long term condition. The main objective of this analysis is to evaluate the possibility of creep rupture, as per the model of Kawamura (1985). On this model, the rupture for undrained creep is associated with the divergent behavior of the evolution displacements with time, while the convergent behavior would indicate consolidation and stabilization. The tendency observed is clearly convergent during and after construction, thus indicating the stabilization condition. The measured maximum horizontal displacements just after the end of construction (140 days) were in the order of 80–97 mm and in the long term (885 days) in the order of 155 mm.

Fig. 10 presents the maximum horizontal displacements normalized as a function of the thickness of the clay level (Y_{max}/D) versus time for the access embankments of the Jitituba River bridge and for the Juturnaíba trial embankment. The tendency observed is in agreement with what is proposed and with the design condition of each embankment. It was also observed that the magnitude of Y_{max}/D for the access embankments of the Jitituba River Bridge is much lower than the relation observed in the experimental Juturnaíba

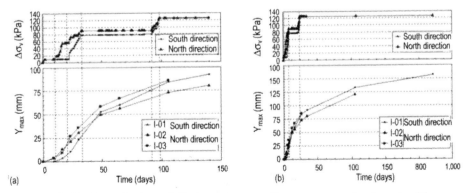

FIG. 9 *Maximum horizontal displacements over time – access embankments of the Jitituba River Bridge: (a) up to 140 days; (b) up to 885 days (Cavalcante et al., 2003).*

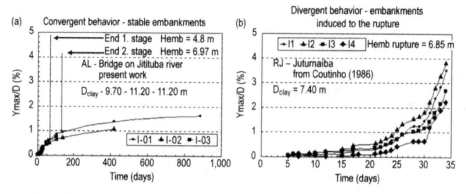

FIG. 10 *Relation Y_{max}/D (horizontal maximum displacements / thickness of the clay layer) with time – access embankments of the Jitituba River Bridge: (a) stable embankments; (b) embankments induced to the rupture (Cavalcante, 2001; Cavalcante et al., 2003)*

embankments induced to rupture. The divergent behavior in Juturnaíba becomes more evident after 30 days (H_{emb} = 5.60 m; FS = 1.31), showing in the inclinometer I-3, that the Y_{max}/D value is around 1.2% (limit of stability) and for the last reading corresponding to H_{emb} just before failure, the Y_{max}/D value is 2.75% (beginning to be unstable).

The divergent and convergent behavior of the Y_{max}/D curve with time directly relate to the velocity of the horizontal deformation. Cavalcante et al. (2003) verified the possibility of evaluating the unstable or stable situation of the foundation soil through the velocity of the horizontal deformation. Fig. 11 shows the results of the relation Y_{max}/D, which presents maximum values between 0.02 and 0.03%/day, for the stable embankment, and between 0.4 and

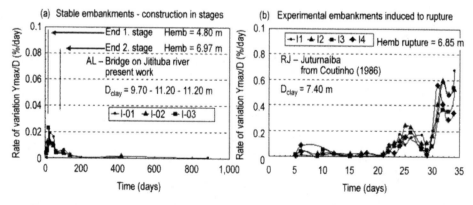

FIG. 11 *Rate of variation of the relation Y_{max}/D (horizontal maximum displacements / thickness of the clay layer) with time – access embankments of the Jitituba River Bridge: (a) Stable embankments; (b) embankments induced to rupture (Cavalcante et al., 2003)*

TABLE 3 SUMMARY OF THE STABILITY CONTROL PROPOSAL FROM HORIZONTAL DISPLACEMENTS

	Analysis methods	Classification
$Y_{max}/D \times$ time	Tendency with time (Kawamura, 1985) Maximum value	Convergent behavior → Stable Divergent behavior → Unstable
	Rate of variation ($\Delta Y_{max}/D)/\Delta t$ (Cavalcante, 2001; Cavalcante et al., 2003)	<<0.2% / day → Stable >0.2% / day → beginning Unstable
	Tendency with the time (Cavalcante, 2001; Cavalcante et al., 2003)	Convergent behavior → Stable Divergent behavior → Unstable
Angular distortion × time	Maximum value Construction end (Ortigão, 1980; Coutinho, 1986)	<3% → Stable >3% → beginning Unstable
	Rate of variation (Almeida et al., 2001)	vd ≥ 1.5%/day → Tendency to be unstable 0.5% ≤ vd ≤ 1.5% → Alert, especial attention vd≤0.5%/day→Stable, to continue monitoring
Displaced vertical volume × Displaced horizontal volume	($\Delta Vv/\Delta Vh$) (Sandroni et al., 2004) (Johnston, 1973)	($\Delta Vv/\Delta Vh > 5$) → Stable (3 < $\Delta Vv/\Delta Vh$ < 5) → Medium (alert) ($\Delta Vv/\Delta Vh$ < 3) → Unstable
$\delta h_{max}/D \times$ FS	(Bourges and Mieussens, 1979; Cavalcante, 2001; Cavalcante et al., 2003)	>1.8% → Unstable =1.0% → Stable (FS~1.5) < 0.8% → Stable (Minimum horizontal displacements)

1.0%/day, for the embankments induced to rupture. In Juturnaíba (Fig. 11b), it is observed that after 30 days a large increment occurs in the rate of variation of the Y_{max}/D value, showing a stable limit value around 0.20%/day. The results were similar for the Sarapuí trial embankments and Juturnaíba Dam. Through the results obtained, the maximum value of the rate of variation of Y_{max}/D in stable condition (FS_{med} = 1.31) was about 10 times greater in the embankment induced to failure in comparison to the embankments constructed to be stable (FS~1.5).

4.2.2 Angular Distortion with Time

Fig. 12 presents the results of maximum angular distortions (10^{-2} radians or %) with time, for the Jitituba embankment. It can be observed that the curve showing a stable maximum value with the time (< 2% for 140 days and 3.55% for 885 days). The convergent stable behavior is similar to the one observed in the analysis of the maximum horizontal displacement.

Ortigão (1980) and Coutinho (1986) in the analysis of the behavior in the experimental embankments of Sarapuí and Juturnaíba found the maximum angular distortions in the rupture to be around 17 and 12%, respectively, thus showing divergent behavior. According to the authors, both embankments stayed in a stable situation during construction while the maximum angular distortion value was lower than 3% (Sarapuí) and 4% (Juturnaíba-Fig. 13). Fig. 14 presents the value of the maximum rate of angular distortion (*vd*) over time for the Jitituba embankment. The maximum value observed was 0.07%/day and this occurred in the first construction period, thus showing a much lower value than the limit vd < 0.5%/day proposed by Almeida et al. (2001) for a stable situation.

Table 4 shows the results of *vd* obtained for the trial Juturnaíba embankment. The values were relatively small and the embankment stable until an embankment height of 5.60 m (FS = 1.31). When the embankment height was increased from 5.60 to 6.10 m, a significant increase in *vd* occurred for

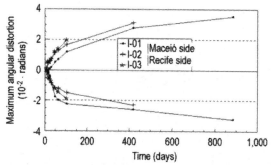

FIG. 12 *Maximum angular distortion with time, I-01, I-02, I-03 (Cavalcante et al., 2003)*

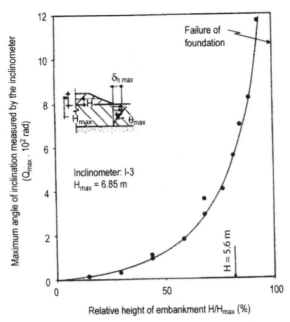

FIG. 13 *Maximum angular distortion with the height of embankment (Coutinho, 1986)*

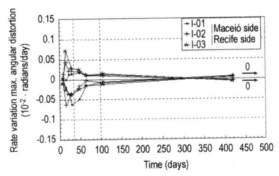

FIG. 14 *Rate of variation of the maximum angular distortion with time (Cavalcante et al., 2003)*

all inclinometers, particularly for the I-2, showing extremely high values for $H_{emb} = 6.40\,m$ and an imminent collapse ($H_{emb} = 6.85\,m$).

4.2.3 Relationship between the Variation of Vertical Volume and the Variation of Horizontal Volume

Sandroni et al. (2004) presented a stability control proposal with analysis and discussion on this topic. The proposed methodology is based on two relations: (a) t (time) vs. Vv/Vh or dVv/dVh; and (b) H (height of embankment) vs. Vh. In the methodology, it is recommended an analysis to be made of the behavior

TABLE 4 ANGULAR DISTORTION RATE JUTURNAÍBA TRIAL EMBANKMENT, vd (COUTINHO, 1986; ALMEIDA ET AL., 2001)

H (m)	I-1	I-2	I-3	I-4	I-5
4.65	0.3	0.6	0.5	0.5	0.5
5.60	0.3	0.6	0.4	0.4	0.4
6.10	1.3	2.3	1.3	1.4	1.6
6.40	2.7	2.5	3.1	1.4	2.4

of all relations proposed to verify the condition of stability. Values of the relations in (a) higher than 5 is a sign of stability. There is an unstable condition (or proximity of failure) when any of the relations Vv/Vh or dVv/dVh present a value less than 3. Another unstable condition is the inclination of the relation H vs. Vh increasing significantly (Fig. 15).

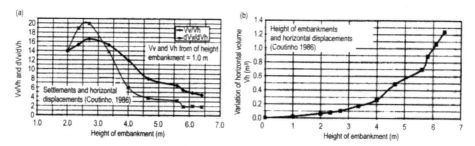

FIG. 15 *Juturnaíba trial embankment: (a) failure around 35 days; (b) evolution of horizontal volume with the width of embankment (Sandroni et al., 2004)*

This proposal was applied to the Jitituba embankment (Cavalcante et al., 2003). The results were in the ranges of 8.4 to 28.0% (South direction) and 4.2 to 28% (North direction), which in general was classified as a stable condition ($\Delta Vv/\Delta Vh > 5$), and only in some intervals, during the first phase of the construction, the embankment presented a situation classified as medium alert ($3 < \Delta Vv/\Delta Vh < 5$). Johnston (1973) observed the range of $3.5-4.2 < Vv/\Delta Vh \leq 20$ as corresponding to a behavior of partial drainage and without failure of embankments; and in the embankment which presented failure, the relation Vv/Vh was in the range of $2.4 - 1.8$.

Fig. 16 shows the results of t (time) vs. Vv/Vh or dVv/dVh and H (height of embankment) vs. Vh obtained from the Juturnaíba trial embankment. Above an embankment height of 5.60 m (FS = 1.3), the behavior changed significantly showing results that correspond to the beginning of a possible process of failure, which occurred shortly afterwards, with H_{emb} = 6.85 m.

4.2.4 Relationship between Horizontal Displacements and Factor of Safety (or embankment height)

Coutinho (1986) in the Juturnaíba trial embankment discussed the use of

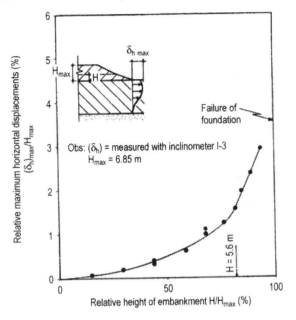

FIG. 16 *Relative horizontal displacements with the relative height of embankment, I-3 (Coutinho, 1986)*

the relation Y_{max}/H_{emb} (%) versus embankment height in the stability control. As to the results, there is a sharp increase after H_{emb} = 5.6 m, and it being recommended that this relation be less than 1.5% to be stable.

Cavalcante et al. (2003) presented and discussed results of the relation H_{embmax}/D (thickness of soft deposit) versus factor of safety (FS) for use as a stability control proposal. Fig. 17 shows the results obtained for some Brazilian embankments on soft clays. It is observed that the tendency of the FS values decreases when the Y_{max}/D increases. For the case of the stable Jitituba embankment, the maximum value of the relation Y_{max}/D for the 1st and 2nd construction stages was 0.54% and 0.32%, respectively.

Considering these cases, the authors proposed values for stability control during the construction period: a) $Y_{max}/D > 1.8\%$ indicates rupture situations (FS = 1.0); b) $Y_{max}/D=1.0\%$ is used for the safety factor generally adopted in practice (FS = 1.5); c) $Y_{max}/D < 0.8\%$ indicates limited horizontal displacement.

Table 5 presents a summary of the results obtained in the analysis of stability control using all proposals from Table 3 for the four cases studied: two embankments constructed to be stable (Jitituba River and Juturnaíba embankments). These results illustrate the stability control proposals.

Ladd (1991) presents and discusses the use and interpretation of field displacements data as a stability control. It is pointed out that this requires experience and judgment and the use of different "graphs", depending on each

TABLE 5 SUMMARY OF THE RESULTS OBTAINED IN THE ANALYSIS OF STABILITY CONTROL

ANALYSIS METHODS	RESULTS / CLASSIFICATION			
	(1)	(2)	(3)	(4)
Tendency over time (Y_{max}/D vs. t) (Kawamura, 1985)	I2 and I4 – Divergent H_{emb}=2.5 m - 0.7-0.9% H_{emb}=2.8 m - 1.5-1.7% **Unstable**	I3 – Divergent H_{emb}=5.6 m - 1.22% **Stable** H_{emb}=6.4 m - 2.75% **Beginning to be Unstable**	I1, I3 and I4 – Convergent H_{emb}=10.7 m - 1.8; 3.4; 3.6% **Stable**	Convergent 1.62% (157 mm) **Stable**
Rate of variation ($\Delta oY'_{max}/D)/\Delta t$ (Cavalcante, 2001; Cavalcante et al., 2003)	I3 and I4 H_{emb}=2.5 m - 0.25%/day H_{emb}=2.8 m - 0.5%/day **Unstable**	I1, I2, I3 and I4 H_{emb}=5.6 m - 0.25%/day **Stable** H_{emb}=6.4 m - 0.5%/day **Beginning to be Unstable**	0.030% / day **Stable**	0.024% / day **Stable**
Angular distortion × Time — Tendency over time (Cavalcante, 2001; Cavalcante et al., 2003)	I3 and I4 – *Divergent*	I3 – *Divergent*		Convergent behavior **Stable**
Construction end (Ortigão, 1980; Coutinho, 1986)	3.0% - **Stable** 15% - **Unstable**	H_{emb}=5.0 m - 3% **Stable** H_{emb}=6.4 m - 12% **Beginning to be Unstable**		<2% **Stable**
Rate of variation (Almeida et al., 2001)	H_{emb}=2.5 m - 1.00%/day **Stable** H_{emb}=2.8 m - 3.5%/day **Unstable**	H_{emb}=5.6 m - 0.4%/day **Stable** H_{emb}=6.4 m - 3.1%/day **Beginning to be Unstable**		0.07% / day **Stable**
Displaced vertical × Displaced horizontal volume ($\Delta Vv/\Delta Vh$) (Sandroni and Lacerda, 2001; Sandroni et al., 2004)	Vv/Vh or $\Delta Vv/\Delta Vh$ > 7% - **Stable** < 3% - **Unstable**	Vv/Vh or $\Delta Vv/\Delta Vh$ > 3% - **Stable** = 2% - **Unstable**		South direction (8.4 to 28.0%) **Stable** North direction (4.2 to 28%) **Medium to Stable**
$\delta h_{max}/D \times f$ (Bourges and Mieussens, 1979; Cavalcante, 2001; Cavalcante et al., 2003)	I3 and I4 – Divergent H_{emb}=2.8 m - 2.7% **Unstable**	I3 – *Divergent* H_{emb}=5.6 m - 1.5% **Stable** H_{emb}=6.4 m - 2.7% **Beginning to be Unstable**	I1, I2 and I3 – Convergent H_{emb}= 6.0 m (1st Stage) 1.93%; 0.90%; 2.09% **Stable**	0.44 to 0.54% **Stable** (MINIMUM HORIZONTAL DISPLACEMENTS)

(1) Sarapuí trial embankment (Ortigão, 1980) (2) Juturnaíba trial embankment (Coutinho, 1986) (3) Juturnaíba Dam (Lucena, 1994; Coutinho et al., 1994) (4) Access embankments of the Jitituba River Bridge (Cavalcante, 2001; Cavalcante et al., 2003).

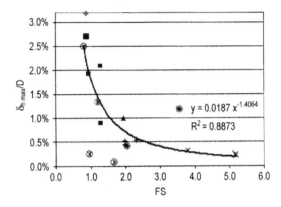

$$y = 0.0187 x^{-1.4064}$$
$$R^2 = 0.8873$$

+ AL – Bridge Jitituba river (1st stage)
× AL – Bridge Jitituba river (2nd stage)
■ RJ – Juturnaíba experimental embankment
■ RJ – Juturnaiba dam (1st stage)
✴ RJ – Sarapuí experimental embankment
▲ RJ – REDUC
✶ SC – BR-101 DNOS chanel
● SC – BR-101 – Inferninho river
○ Geogrids

FIG. 17 *Relation Y_{max}/D vs. FS: Brazilian embankments (Cavalcante et al., 2003)*

problem. The authors recommend the use of many of the graphs/proposals shown in this paper. It is also to be remembered that the behavior of soft clay foundation can be brittle or ductile. In soils with brittle behavior, such as sensitive clays, the rupture can be abrupt and difficult to anticipate. In soils with ductile behavior, the process tends to be more gradual and to make it more possible to give sufficient anticipated signals.

5   FINAL COMMENTS AND CONCLUSIONS

This paper presents and discusses results of stability analysis and stability control. Research and practical cases were used with emphasis on the study performed in the Juturnaíba trial embankment. Evaluation of mobilized shear strength Su in an embankment constructed in one stage obtained by some approaches was presented and discussed. In the total stability analysis, the cases of failure in Brazilian embankments on soft clays (the exception is the Juturnaíba trial embankment) show the need to apply the Bjerrum (1973) correction factor to undrained strength from field vane tests. The presence of organic soil layers in the Juturnaíba foundation with strong drainage and deformation/increases of stress during the construction seems to be one possible reason for the "different" behavior. Effective stress stability analysis considering normally consolidated effective stress parameters of strength presented satisfactory results, particularly when cracking of the embankment was considered to prompt the failure.

Stability control is one of the important steps in the design and construction of embankment on soft soils. This can be realized using the measures of displacements, deformations or pore pressures. Proposals were presented and analyzed, particularly for horizontal displacements, and show potential for use in practical work, depending on each problem. Due to the limits of each proposal and the many variables involved in the process, it is recommended more than one proposal be used to have more confidence in the decision. The joint results of analysis and control of stability showed the importance of having FS > 1.3 to guarantee adequate behavior and security.

ACKNOWLEDGEMENTS

An acknowledgement is due to CNPq and CAPES for their financial support. Special thanks to Maria Helena Lucena and Sarita Cavalcante who contributed data used in this paper, and to Gusmão Associated Engineer for entering into a partnership with GEGEP that proved to be of great benefit.

REFERENCES

Almeida, M. S. S., Oliveira, J. R. M. S. and Spotti, A. P. (2001). *Previsão e Desempenho de Aterro Sobre Solos Moles: Estabilidade, Recalques e Análises Numéricas.* Encontro das Argilas Moles Brasileiras, COPPE/UFRJ e ABMS, pp. 166-191, Rio de Janeiro.

Azzouz, A. S., Baligh, M. and Ladd, C. (1983). *Corrected Field Vane Strength for Embankment Design.* ASCE JGE, Vol. 109 (5), pp. 730-734.

Bello, M. I. M. C. (2004). *Estudo de Ruptura em Aterros Sobre Solos Moles - Aterro do Galpão localizado na Br-101-PE.* M.Sc. thesis, Federal University of Pernambuco, Brazil, in Portuguese.

Bello, M. I. M. C.; Coutinho, R. Q. and Gusmão, A. D. (2006). *Estudo de ruptura de um aterro sobre solos moles localizado em Recife, Pernambuco.* In: XIII COBRAMSEG.

Bjerrum, L. (1973). *Problems of Soil Mechanics and Construction of Soft Clays and Structurally Unstable Soils.* 8th ICSMFE, (2), pp. 111-159.

Bourges, F. and Miussens, C. (1979). *Déplacement Latéraux à Proximité des Remblais Sur Sols Compressibles – Méthod de Prévision.* Bull. Liaison Labo. P. et Ch., n. 101, pp. 73-100.

Cavalcante, S. P. P. (2001). *Análise de Comportamento de Aterros sobre Solos Moles - Aterros de Encontro da Ponte sobre o Rio Jitituba- AL.* M.Sc. thesis, Federal University of Pernambuco, Brazil, in Portuguese.

Cavalcante, S. P. P., Coutinho, R. Q. and Gusmão, A. D. (2003). *Evaluation and Control of Stability in Field of Embankments on Soft Soils.* XII PCSMGE, Cambridge, MA, pp. 2649-2660.

Coutinho, R. Q. (1986). *Aterro Experimental Instrumentado Levado à Ruptura Sobre Solos Orgânicos - Argilas Moles da Barragem de Juturnaíba.* D.Sc. thesis, COPPE / UFRJ, Brazil, in Portuguese.

Coutinho, R. Q., Almeida, M. S. S. and Borges, J. B. (1994). *Analysis of The Juturnaíba Embankment Dam Built on Soft Clay.* ASCE, Geotechnical Special Publication No. 40, Vol. 1, pp. 348-363.

Coutinho, R. Q. and Lacerda, W. A. (1987). *Characterization-consolidation of Juturnaíba organic clays.* Proc. ISGESS, Mexico, Vol. (1), pp. 17-24.

Coutinho, R. Q. and Lacerda, W. A. (1989). *Strength characteristics of Juturnaíba organic clays.* Proc. 12th ICSMFE, RJ, V.(3), pp. 1731-1734.

Coutinho, R. Q. and Bello, M. I. M. (2005). *Geotecnia do Nordeste-* Livro ABMS / Núcleo Nordeste. (3), pp. 1-26.

Johnston, I. W. (1973). *Discussion - Session 4. In Field Instrumentation in Geot. Eng,* Halsted Press Book, John Wiley, New York, pp.700-702.

Kawamura, K. (1985). *Methodology for landslide prediction.* Proc. XI ICSMFE, San Francisco, V. 3. pp. 1155-1158.

Ladd, C. C. (1991). *Stability evaluation during staged construction.* JGE, ASCE, Reston, VA, 117 (4). 540-615.

Leroueil, S. and Rowe, R. K. (2000). *Embankments over Soft Soil and Peat.* Geotechnical and Geoenviromental Engineering. Handbook. Edited by R. Kerry Rowe. (16), pp.463-498.

Magnani de Oliveira, H. (2006). *Comportamento de aterros reforçados sobre solos moles levados à ruptura.* D. Sc. Thesis. COPPE/UFRJ. Portuguese.

Mesri, G. (1975). *Discussion: New Design Procedure for Stability of Soft Clays.* ASCE JGE, Division 101(GT4): 409-412.

NAVFAC DM-7. (1971). *Soil Mechanics, Foundations and Earth Structures Design Manual,* U.S. Dep. of the Navy, Washington, D. C. 20390.

Ortigão, J. A. R. (1980). *Aterro Experimental Levado à Ruptura sobre Argila cinza do Rio de Janeiro.* D.Sc. thesis, COPPE / UFRJ, in Portuguese.

Pillot, G. and Moreau, M. (1973). *La Stabitité des Remblains sur Soil Mour - Abaqus de Calcul,* Editions Eyrolles, Paris, 151p.

Pinto, C. S. (1966). *Capacidade de Carga de Argilas com Coesão Linearmente crescente com a Profundidade,* in: Jornal de Solos, 3(1) pp. 21-44. São Paulo, Brazil.

Sandroni, S. S. (2006). *Sobre a prática brasileira de projeto geotécnico de aterros rodoviários em terrenos com solos muito moles.* In: Congresso Luso-Brasileiro, Curitiba.

Sandroni, S. S. and Lacerda, W. (2001). *Discussão sobre Controle de Estabilidade em Aterros sobre Solos Moles.* Encontro Propriedades das Argilas Moles Naturais Brasileiras, COPPE/UFRJ – ABMS.

Sandroni, S., Willy, A. L. and Brandt, J. R. T. (2004). *Método dos volumes para Controle de Campo da Estabilidade de Aterros sobre Argilas Moles.* Solos & Rochas, ABMS, São Paulo, Brazil, Vol. 27(1), pp. 37-57.

Tavenas, F. and Leroueil, S. (1980). *The Behavior of Embankments on Clay Foundations.* Canadian Geotechnical Journal, Vol. 47(2), pp.236-260.

Terzaghi, K. (1943). *Theoretical Soil Mechanics*, Wiley, New York.

Tall buildings in the city of Santos: old shallow foundations unexpected behavior and the newly employed long piles

Faiçal Massad
Escola Politécnica da Universidade de São Paulo, Brazil

KEYWORDS tall building, unexpected behavior, shallow foundation, instrumented piles

ABSTRACT This paper deals with some unexpected behaviors on the foundations of tall buildings in the cities of Santos and São Vicente, including both the old shallow foundations and the newly set long piles. The findings referred to herein are explained based on the geological history of Santos Coastal Plains marine clays. Such plains have been gradually unveiled since the 1980s. First, it is now known that those clays are not as bad as it was first thought: there are occurrences of medium to hard clays next to the more commonly found very soft to soft clays. Moreover, these marine sediments were submitted to loads from sea level variations and dune action, for better or for worse, with OCR ranging from 1.1 to above 2.5; more than 12 experimental fills settled in a wide range of values and of velocities. Third, the erratic over-consolidation of such soft clays in the city of Santos, underlying a thick layer of compact sand, explains some of the anomalous behavior of tall buildings rested on shallow foundations, as: a) the scattering in the primary settlements and in the secondary consolidation rates; b) the greater settlement in the least loaded corner; and c) the slight to moderate inclination of some specific buildings with no influence of or being influenced by other surrounding buildings. Last, but not least, long steel piles in the city of Santos behave like piled raft due to the presence of the upper thick layer of compact sand overlying soft clays.

1 INTRODUCTION

Sedimentary clays of the Brazilian coastal plains, particularly in Santos, for a long time were considered to be soft, normally consolidated clays formed during a single depositional cycle, without erosion (Pacheco Silva, 1953). This conception lasted for many decades, despite the fact that since the end of the 1940s it has been known that medium to hard clays – sometimes highly over consolidated – apparently of the same origin might occur deeper in the subsoil. In the 1970s, the so called soft clays revealed themselves to be lightly overconsolidated, due to a reason unknown at that time.

In the 1980s a discovery was made that relative sea-level oscillation (shore level displacement) during the Quaternary period was the main cause of sedimentation on the Santos Coastal Plain ("Baixada Santista"). At least two depositional cycles occurred, with an intermediate erosive process between them which is associated with transgression episodes of high sea-levels, that were the origin of two kinds of clayey sediments, each with properties very different from the other.

This paper deals with these and other unexpected findings related to the behavior of tall buildings, built over shallow foundations in that area over the last 60 years in the light of the geological history that has been gradually unveiled. Some attention will be given to the performance of long steel piles recently installed and instrumented in the city of Santos, which in turn also behaved in an unexpected manner.

2   THE CLAYS OF SANTOS ARE NOT THAT BAD

In 1949 many borings made for the construction of the Anchieta Highway Casqueiro Bridge revealed the occurrence of two layers of soil (Fig. 1). The upper one was of soft clay, usually found in the area. The lower layer, of medium clay, 25 m to 30 m deep over consolidated ($\sigma'_p \sim 360\,\text{kPa}$) to which scientists had no explanation at that time. It is worth mentioning the deep layer of black clay, very hard (SPT > 25), found in many places as shown in Fig. 1.

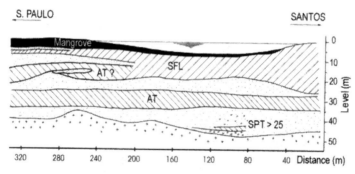

FIG. 1 *Geological section – Casqueiro Bridge*

However, the most surprising case occurred in 1954-1955, during the foundation of Building I, in the city of São Vicente (Fig. 2). The subsoil consisted of an upper compact layer of sand, 12.6 m thick, overlying 30 m of hard clay, 5 to 10 blows, highly over consolidated ($\sigma'_p \cong 600$ to $700\,\text{kPa}$). The surrounding soil was very different, though: soft clay, 1 to 2 blows, with OCR~1. Again, no justification was given at that time.

The explanation for most of theses findings came in the 1980s. Many geological evidences show that the sedimentary clays of the Santos Coastal Plain ("Baixada Santista") were formed over, at least, two Quaternary depositional cycles, with an intermediate erosive process (Suguio and Martin, 1981). This gave origin to the Pleistocene Clays and to the Holocene Clays. The former one, also called Transitional Clays (AT), deposited 100,000 to 120,000 years BP (Before Present) are medium to stiff clays. The latter ones are very soft clays and include: a) SFL clays (from Sediments-Fluvial-Lagoon-Bay) originated

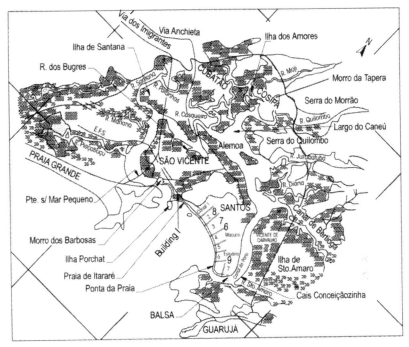

FIG. 2 *Islands of Santos and São Vicente – Baixada Santista*

some 10,000 years BP by sedimentation where Pleistocene sediments eroded; and b) mangrove sediments that are still forming. As far as the very hard clays, it is reasonable to ask: are such soils the remains of two transgressions that took place in the south coast of Brazil instead of just the last one?

Table 1 presents the main properties of Marine Clays. Their index properties are almost the same, but they differ in "state properties", such as undrained shear strength, void ratio and SPT.

TABLE 1 SOME PROPERTIES OF MARINE HOLOCENE AND PLEISTOCENE CLAYS

Item	Differences			Item	Similarities		
	Mangrove	SFL	AT		Mangrove	SFL	AT
Depth(m)	≤ 5	≤ 50	20-45	Depth(m)	≤ 5	≤ 50	20-45
SPT	0	0-4	5-25	γ_n(kN/m^3)	13.0	13.5-16.3	15.0-16.3
B_q	—	0.4-0.9	−0.1-0.2	%<5μ	—	20-90	20-70
q_t(MPa)	—	0.5-1.5	1.5-2.0	w_L	40-150	40-150	40-150
E	>4	2-4	<2	I_P	30-90	20-90	40-90
σ'_p(kPa)	<30	30-200	200-700	$C_c/(1+e_o)$	0.36	0.43	0.39
OCR	1	1.1-2.5	>2.5	Cr/Cc(%)	12	8-12	9
s_u(kPa)	3	10-60	>100	R_f(%)	—	1.5-4.0	1.5-2.0

Caption: see appended List of Symbols

3 Overconsolidation of Clays

On one hand, medium to hard Transitional Clays (AT) were pre-consolidated by a sea-level lowering of about 110 m at the peak of the last glaciation (15,000 years BP). This fact is confirmed by consolidation tests on undisturbed samples taken from 6 boreholes (Massad, 1987, 2004, 2009c).

On the other hand, the SFL clays are lightly over consolidated due to such occurrences as short negative sea-level oscillations (i.e., bellow what the present sea level is), dune action and aging effects (Massad, 1987, 2004, 2009c). For each of about 20 stress history profiles (SHP), from odometer tests and SCPTU, like those of Figs. 3 and 4, the following relationship holds:

$$\sigma'_p - \sigma'_{vo} = \text{constant} \qquad (1)$$

Fig. 5 illustrates the validity of Eq. 1 for various SCPTUs carried out in two sites of the Santos Coastal Plain. Table 2 shows values of the constant of Eq. 1 for the Classes in which the Holocene SFL Clays may be classified according to both the type of outcropping layer and the prevailing pre-consolidation

TABLE 2 CLASSES OF HOLOCENE CLAYS (SFL). SANTOS COASTAL PLAIN ("BAIXADA SANTISTA")

Class	Clay Location	Over consolidation Mechanism (Tests)	SPT	OCR	Site	$\sigma'_p - \sigma'_{vo}$ (kPa)
1	Outcropping	Neg. Sea-level Osc. (12 SHP)	0	1.3-2.0	Santos Coastal Plain	20-30
2		Dune Action (2 SHP+10 SCPTU)	1-4	>2.0	Santo Amaro Island	50-120
3	Beneath 8-12 m of	Neg. Sea-level Osc. (4 SHP+1 SCPTU)	1-4	1.0-1.3	City of Santos	15-30
4	sand layer	Dune Action (2 SHP)	1-4	>1.4	City of Santos	40-80

Caption: see appended List of Symbols

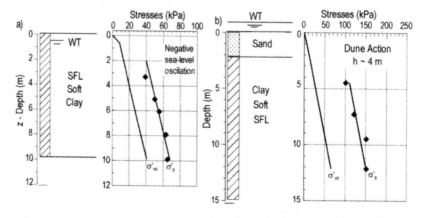

FIG. 3 *Odometer tests: a) Piaçaguera-Guarujá Highway and b) Santo Amaro Island*

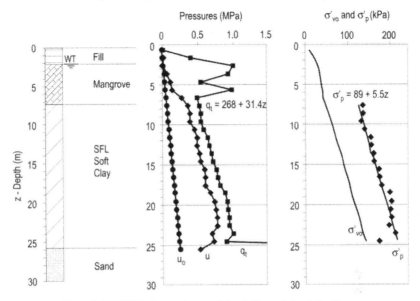

FIG. 4 *SCPTU-9, Santo Amaro Island, Conceiçãõzinha Quay*

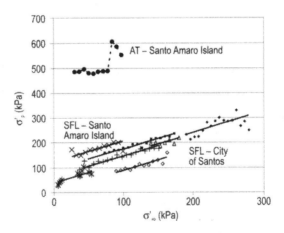

FIG. 5 σ'_p *from SCPTUs as a function of the* σ'_{vo}

mechanism. While Classes 1 and 2 predominate mainly in the inner parts of the Santos Coastal Plain, Classes 3 and 4 occur in the City of Santos (Massad, 1987, 2004, 2009c). The aging effect may result in a rate $\Delta\sigma'_p/\Delta\sigma'_{vo} = 1.15$ instead of 1 of Eq. 1; in this paper it will be neglected. These findings have a practical consequence: more than 12 experimental fills settled in a wide range of values ($\varepsilon_{EOP} = 1$ to 12%) and velocities ($C_v \sim 3.10^{-3}$ to 5.10^{-2} cm^2/s), depending on OCR (Massad, 1987, 2009c).

Its is worth mentioning that the linear increasing in q_t with depths (see Fig. 4) with rate b observed in all SCPTUs may be used (Massad, 2009a) to estimate $N_{\sigma t}$ of the equation (Kulhaway and Mayne 1990):

$$\sigma'_p = \frac{q_t - \sigma_{vo}}{N_{\sigma t}} \quad \text{with} \quad N_{\sigma t} = \frac{b - \gamma_n}{\gamma'} \tag{2}$$

The application of Eq. 2 to a SCPTU near Barnabé Island resulted in $N_{\sigma t} = 3.9$ and the σ'_p (full line) of Fig. 6, which also shows data of odometer tests (dashed line is the average) (Andrade, 2009). The agreement between the two different measurements is good.

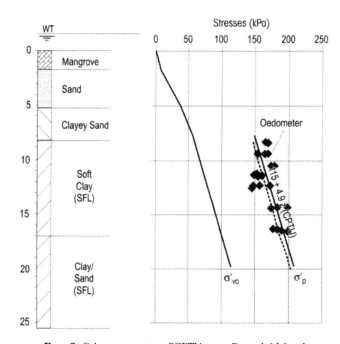

FIG. 6 *Odometer tests × SCPTU, near Barnabé Island*

4   SANTOS BUILDINGS BUILT ON SHALLOW FOUNDATIONS

The soft SFL clays of the cities of Santos and São Vicente were deposited more than 7,000 years BP, probably over AT sediments (Fig. 7). They are overlaid by a 8-12 m thick sand deposit, originated from the displacement of barrier islands developed over periods of land submergence. Such barrier-islands formed lagoons on their backside that lasted partially isolated for relatively long periods of – almost – stable sea-level. With the subsequent rapid lowering of the sea-level (periods of land emersion), the barrier-islands displaced toward the continent completely isolated these lagoons from the open sea and caused their desiccation. Later on, intra-lagoon river deltas were formed

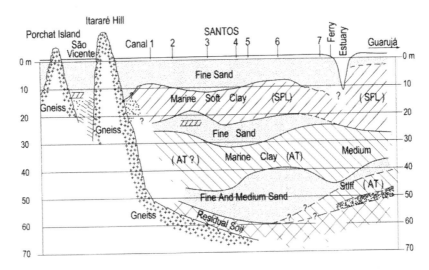

FIG. 7 *Geological Section. Santos and São Vicente Shoreline*
(Adapted from Teixeira, 1994)

in quiet seawaters originating the Santos Coastal Plain. Eolic deposits have always been present in the area.

Santos has a population of over 500,000 people and has the biggest harbor in Latin America. From the 1940s up to the 1970s a booming tourist industry led to the construction of many tall buildings – of up to 18 floors or even higher – along the beach shore. Shallow foundations built on the upper layer (Fig. 7) of medium to compact sand supported them. In general, the maximum settlements ranged from 0.4 m to 1.2 m (Teixeira, 1994). In some cases with an unexpected scatter of values: buildings with the same height and same soft layer thickness settled differently, in proportions as 1:3. Moreover, today there are about 100 leaning buildings, and in one extreme case a 2.2° tilted building was straighten up as reported by Maffei et al. (2001). Since the 1950s the cause of such leanings has been attributed to: a) the loaded area "T" and "L" shaped forms (see Fig. 8a) not recommended; b) highly non uniform loads;

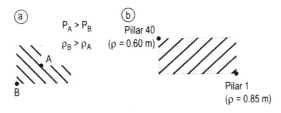

FIG. 8 *Anomalous behavior: (a) "L" shaped loaded area; and (b) leaning of SA Building (P is the load and ρ is the settlement) (Teixeira, 1960b)*

and c) the construction of nearby buildings, as close to each other as 4 m to 10 m (Teixeira, 1994). Nevertheless, some tiltings occurred without a logical explanation, as it is the case of the SA Building (Fig. 8b).

Table 3 presents the results of settlement analyses carried out on 10 buildings 8 to 15 floors high (N), most of them continuously measured over a period of 5 to 10 years. They were built over the years of 1947-1954, except for the UNISANTA Building, erected at the end of the 1990s. As an illustration of the type of analysis carried out, Figs. 9a and 10a show the settlements of the most settled column of 2 building, and Figs. 9b and 10b the corresponding values of the settlement velocities.

In these figures the theoretical values were computed; firstly by applying the Asaoka's and Baguelin's (1999) Methods to the measured settlements; and secondly, by using Olson's Consolidation Theory (Olson, 1977), considering the construction time given in Table 3. The agreement along the primary consolidation time is remarkable. Figs. 9c and 10c present the product $v.t$, which is constant along the secondary range and allows the estimation of $C_{\alpha\varepsilon}$ (Massad, 2005, 2009c). Theoretically, the secondary consolidation starts at $U = 85$ to 95%; this means that Hypothesis B of Jamiolkowski et al. (1985) holds true for SFL Clays. Moreover, buildings with almost the same number of floors (N) and soft clay layer thickness (H) showed very different EOP settlements, and the final strain at EOP ranged between 1 and 5%.

Fig. 11a presents the correlation between the settlements of the most settled columns, the maximum applied loads, and OCR. Full lines were determined using mean soil parameters (24 h oedometer tests), like those given in Table 1, without any correction despite ε'_{EOP} low values (Table 3) – probably

TABLE 3 GENERAL DATA AND RESULTS OF SETTLEMENT ANALYSIS OF THE CRITICAL COLUMNS

Building	t_c (days)	N	H (m)	Primary		Secondary		ε'_{EOP} $10^{-11}/s$	Reference
				ρ_{EOP} (mm)	ε'_{EOP} (%)	t_{sec} (days)	U_{sec} (%)		
B (Santos)	650	15	8.0	258	3.2	900	95	21	Machado, 1961
C (Santos)	450	12	12.0	345	2.9	950	91	16	Machado, 1961
D (Santos)	450	12	12.0	315	2.6	1,100	93	13	Machado, 1961
IA (Santos)	340	8	13.5	121	0.9	1,000	95	5	Teixeira, 1960a
IB (Santos)	400	15	13.5	215	1.6	1,000	94	9	Teixeira, 1960a
SC (Santos)	400	14	15.0	237	1.6	1,200	95	6	Teixeira, 1960a
SA (Santos)	400	15	15.0	754	5.0	1,800	83	13	Teixeira, 1960a
U (Santos)	340	10	16.0	436	2.7	1,700	85	7	Teixeira, 1960a
UNISANTA (Santos)	100	7(10)	16.0	140	0.9	—	—	—	Gonçalves et al., 2002
I (S.Vicente)	420	13	30.5	12	0.4	1,600	100	—	Teixeira, 1960b

Caption: see appended List of Symbols

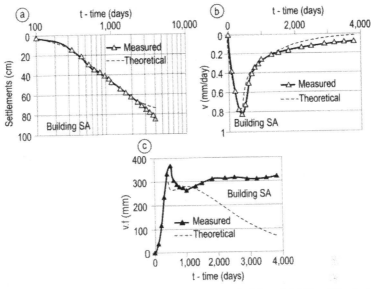

FIG. 9 *Settlements and Settlement Velocities (v) – Building S A (City of Santos)*

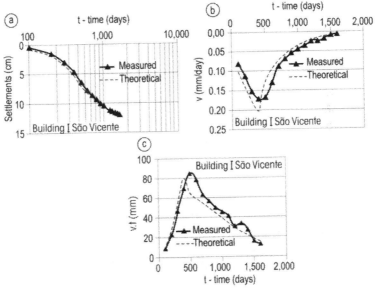

FIG. 10 *Settlements and Velocities (v) – Building I (City of São Vicente)*

due to some sample disturbance (Leroueil, 1996). The numbers associated to the letters identifying the buildings are the OCR measured in each site. These results give a fair explanation of the scatter of EOP settlements mentioned above: they depend on the preconsolidation mechanism (sea level oscillation or dune action) and as a consequence on the class 3 or 4 of Table 2. It is

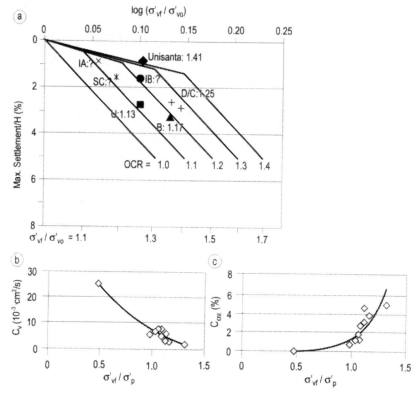

FIG. 11 *a) Max. EOP settlements of Santos Buildings; b) C_v and c) $C_{\alpha\varepsilon}$ as a function of maximum load and OCR*

worth mentioning that the OCR = 1.41 associated to UNISANTA Building is a consequence of dune action, class 4 of Table 2.

The case of Building I in São Vicente is an exception to the rule. As it's been mentioned, it rested on compact sand overlaying a layer of a highly over consolidated stiff clay, with OCR = 2.5 (see Table 3). Due to this fact, the theoretical curves of Fig. 10 fitted remarkably well to the measured values, even for the product $v.t$ (Fig. 11c), with a very small secondary compression – which is consistent with its OCR value.

From Fig. 11b Santos Buildings, C_v ranges from 3×10^{-3} to 8×10^{-3} cm²/s, the same order of magnitude as those of SCPTU (Massad, 2004) and also of those published by Teixeira (1994). As far as the São Vicente Building I is concerned, this figure is as high as 25.3×10^{-3} cm²/s reflecting again the higher OCR value of the AT (Pleistocene clay). On the other hand, $C_{\alpha\varepsilon}$ varies roughly between 1 and 5% (Fig. 11c), averaging 2.5%, for the soft marine SFL clays (Holocene) and is as low as 0.09% for the medium to stiff AT (Pleistocene clay) of Building I.

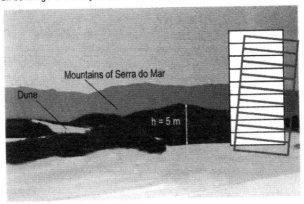

Fig. 12 *Dune at Praia Grande (Massad, 2009c, adapted from Rodrigues, 1965)*

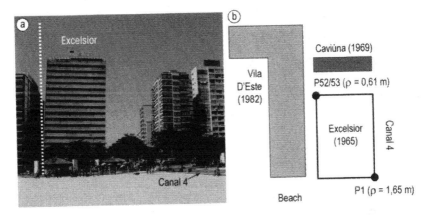

Photo 1 *Excelsior Buiding: a) leaning towards Canal 4; b) Year of construction and settlements (ρ) at corners (Gonçalves, 2006)*

There are many evidences of dune action in Santos plain as well as in its vicinity. In the 1950s, dunes 1 to 5 m in height were still seen along a beach close to the city of Santos. The maximum pressure they exerted could be of the same order of magnitude as those of 2 to 10 floor buildings! This fact could explain the settlement scattering mentioned above. Besides, the non-uniform pressure they exerted may be the cause of the leaning of isolated buildings, as Figures 12 and Photo 1 illustrate.

5   EVIDENCES OF A "PILED-RAFT EFFECT" IN LONG PILES

The use of piles in Baixada Santista is an old practice (Golombek, 1965; Teixeira, 1988; Aoki and Angelino, 1988), keenly related to oil tank and silo constructions in areas with SFL clays of classes 1 and 3 of Table 2. In the city of Santos, with SFL clays of classes 2 and 4, the difficulty to drive piles through the upper compact sand layer (Fig. 7) and the costs involved restricted severely their usage for many years. Alonso and Aoki (1988) used the high pressure water jetting technique to overcome the upper sand layer. They also proposed to embed shorter floating piles in the Pleistocene sand layer above the AT, based on the knowledge of side friction on SFL and OCR of the Pleistocene clays (AT) (Massad, 1988).

Recently, Falconi and Perez (2008) reported the use of long steel H-piles to support tall buildings in City of Santos. Their relatively small cross-sectional areas combined with their high strengths made penetration through the upper dense layer easier. Welding four segments of decreasing cross-sectional areas gave to each pile a slight step-tapered form, thus reducing costs.

Four of such piles were submitted to static vertical loading tests; two of them were instrumented with strain gages installed along the pile length. The building locations are shown in Fig. 2 and Table 4 discloses some data about the piles. Details and more information about the loading tests may be found in Massad (2009b). Below, only instrumented piles will be analysed.

The instrumented pile of Building 8 provided the most complete set of information. Table 5 shows schematically the subsoil profile at the site with SPT data. Strain measurements at distinct depths allow for preparing the plots of Figs. 13a, 13b and 14a. The following conclusions may be drawn: a) the pile behaved as a floating pile; b) the total shaft resistance equals 3,510 kN; c) the unit shaft resistances reached rapidly their maximum values in all levels, except in the upper sand layer (compare y_1 values of Table 5); and d) this

TABLE 4 DATA ON BUILDINGS AND PILES

Building	N	h_c (m)	S_{max}/S_{min} (cm^2)	Working Load (kN)	Set (mm)	Rebound (mm)
6	17	49.0	124/100	2050	69	18
7	20	47.5	136/100	2310	50	18
8	to	51.0	136/100	2320	35	15
9	24	51.0	109/102	1650	45	20

Caption: see appended list

TABLE 5 SOIL PARAMETERS– BUILDING 8, CITY OF SANTOS

z (m)	Soil layer	SPT	f_{max} (kPa)	y_1 (mm)
1 to 12	Upper sand	10 to 30	60	36
12 to 17	SFL clay	1 to 2	20	4
17 to 22	SFL sand (?)	15 to 50	20	4
22 to 33	SFL clay	2 to 5	25	5
33 to 43	AT	4 to 5	65	4
43 to 50	Pleistoc. sediment	5 to 22	130	3
>50	Residual soil	~ 15	—	—

Caption: see appended list

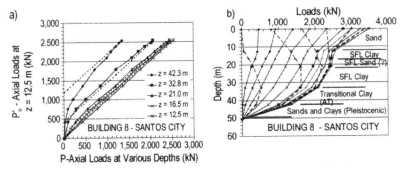

FIG. 13 *a)* P'_o *versus* $P = E \cdot S \cdot \varepsilon$; *b)* Load transfer diagram

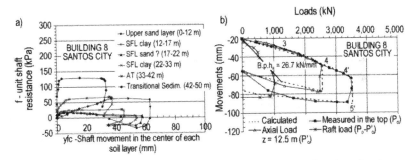

FIG. 14 *a)* Load transfer functions; *b)* Loads-movements (y_o, y'_o and y_{fc})

upper layer exerted a pressure on top of the underlain soft SFL clay, drastically reducing its unit shaft resistance.

The last conclusion suggests the existence of a bulb (an enlargement) in the upper sand layer; or, in other words, the pile behaved as a piled raft. The base of the raft would have to be positioned at the interface between the upper sand and the underlying soft SFL clay, at a depth of 12.5 m. This conjecture justifies the use in Fig. 13a of P'_o, the axial load at depth 12.5 m.

The values of the maximum unit shaft resistances of each soil layer are shown in Table 5. These figures are in agreement with those measured in other Baixada Santista sites (Massad, 1988, 2009c).

Fig. 14b shows the load-movement curves related to: a) the top (P_o vs. y_o); b) the raft (($P_o - P'_o$) vs y_{fc}); and c) the pile of the piled raft (P'_o vs. y'_o). Fig. 15a is analogous to Fig. 14b, but all loads are referred to the top movement. As P'_o is equal to the total shaft resistance between 12.5 and 51 m (pile toe), it can be concluded that the unit shaft resistances reached their maximum values simultaneously, either in the upper sand layer and in the lower layers, due to pile relative high compressibility.

Based on a mathematical model developed by Massad (2009b), Fig. 15b was prepared assuming that the unit shaft resistance of the upper sand layer reached its maximum value for $y_{fc} > 60$ mm.

In this case, the curve P_o vs. y_o would extend to point 6', passing through points 4' (ultimate shaft resistance of the "fictitious piled raft" pile was reached) and 5' (ultimate "raft" load was reached). The straight line 4'-5' represents the remaining shaft resistance of the upper sand layer, or the equivalent load passed on by the "fictitious raft" alone. The equation of this line is (Massad, 2009b):

$$P_o = d_1 + d_2 \cdot y_o \qquad (3)$$

The constants d_1 and d_2 are computed using the following relations:

$$\frac{1}{d_2} = \frac{1}{B.p.h_a} + \frac{1}{4K_{ra}} \qquad (4)$$

$$P'_{or} = \frac{d_1}{1 - d_2/(2K_{ra})} \qquad (5)$$

where: a) $B.p.h_a$ is the "fictitious raft" stiffness; b) K_{ra} is the stiffness of the H-Pile segment embedded in the upper sand layer; and c) P'_{or} is the ultimate total load acting on the fictitious pile of the "piled raft". If line 4'-5' is known, than Eq. 5 or the simple graphical construction of Fig. 15b allows the estimation of P'_{or}.

The H-Pile of Building 6 was also instrumented, but with a restriction in number and positions of strain gages (see Fig. 16a). Nevertheless, the strain measurements lead to the following conclusion (Massad, 2009b): a) the pile behaved as a floating pile; b) the total shaft resistance was greater than 3,309 kN and the ultimate toe load was roughly 190 kN; c) the unit shaft re-sistances reachedrapidly their maximum values in all levels, except in the 3 upper layers; and d) after unloading, a residual load of 60 kN was locked in the toe.

The piled raft effect is evidenced in the graphical construction (Massad, 2009b) of Fig. 16b. It can be seen that the ultimate total load (P'_{or}) acting in

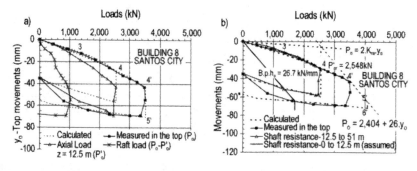

FIG. 15 a) Loads-top movements (y_o); b) Simulation considering higher shaft resistance in the upper sand layer

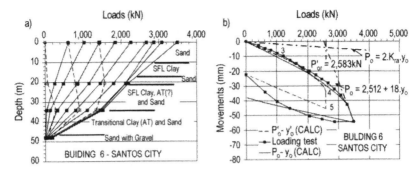

FIG. 16 *a) Load transfer diagram; b) Loads-movements*

the pile of the "fictitious piled raft" is at the order of 2,583 kN and, consequently, the total shaft resistance in the upper sand layer alone is greater than $3,309 - 2,583 = 726$ kN.

6   CONCLUSIONS

The paper presented an explanation for some unexpected behavior of building foundations in the cities of Santos and São Vicente.

I. Old shallow foundations

For tall buildings rested on shallow foundations the erratic over-consolidation of the soft clays of the city of Santos, due to negative sea-level oscillation or dune action, explains the scattering in the settlements and their velocities – both primary and secondary. For Building I in São Vicente, rested on a 30.5 m of Pleistocene stiff clay, primary consolidation predominated; a figure of only 0.09% was guessed for the Coefficient of Secondary Compression ($C_{\alpha\varepsilon}$).

Besides this, the non-uniform pressure exerted by the dunes may respond for both the leaning of isolated buildings without the influence of nearby buildings, and the greater settlement in the least loaded corner of some buildings.

II. Newly employed long piles

For long steel piles set in the city of Santos, the unit shaft resistances reached their maximum values after shaft movements of a few mm – except for the upper sand layer, that required tens of mm. This fact suggested a behavior like a pile toe or an enlargement (bulb) in the upper bearing sand layer.

This piled raft effect brought about a reduction in the unit shaft resistance of the soft SFL clay immediately bellow the upper sand layer. If the resistance of this layer is very high, negative skin friction could develop in the soft clay.

ACKNOWLEDGEMENTS

The author is indebted to GERDAU AÇO MINAS SA for the opportunity to analyze the data of two instrumented H-piles installed in the city of Santos. He also displays his gratitude to the support of the University of São Paulo Politechnic School, Brazil.

LIST OF SYMBOLS

b	Rate of Increase of q_t with Depth (z)
B	Parameter of the 1st Cambefort Law
BP	Before Present
B_q	SCPTU pore pressure coefficient
C_c, C_r	Compression and Recompression Indexes
C_v	Coefficient of Primary Consolidation
$C_{\alpha\varepsilon}$	Coefficient of Secondary Consolidation
e	Void ratio
EOP	End of Primary
E	Young's Modulus of Pile Material
f	Unit Shaft Resistance
f_{max}	Ultimate Unit Shaft Resistance
H	Clay layer thickness
h_a, h_c	Pile Length in the Upper Sand Layer; Pile Embedment
I_P	Plasticity Index
K_{ra}	Stiffness of the H-Pile segment, in the upper sand layer
N; $N_{\sigma\tau}$	Number of floors in the Building; Maynes' Empirical Factor
OCR	Over Consolidation Ratio
p	Perimeter of the H-Piles
P	Axial Load along Pile Length or Load in Columns
P_o	Pile Top Load
P'_o	Top Load of the Pile of the "Fictitious Piled Raft"
P'_{or}	Ultimate Value of P'_o
q_t	Corrected Cone (CPTU) Point Resistance
R_f	SCPTU Friction Ratio
SCPTU	Static Cone Penetration Test with u measurement
SHP	Stress History Profiles from Odometer Tests
S	Cross Sectional Area of the Pile
sec	Secondary
S_{max}; S_{min}	Maximum and Minimum Cross Sectional Area of the Pile
s_u	Undrained shear strength
t; t_c	Time, Construction time
u	Pore Pressure
U; T	Degree of Consolidation and Time Factor
v	Settlement Velocity

y_o	Movement of Pile Top
y'_o	Movement of the Pile Top of the "Fictitious Piled Raft"
y_1	Parameter of the 1st Cambefort Law
y_f	Movement of a Point Along the Shaft
y_{fc}	Value of y_f in the Center of a Soil Layer
z	Depth
w_L	Liquid Limit
γ_n, γ'	Natural and Effective Specific Unit Weight
$\varepsilon; \varepsilon_{EOP}$	Strain and EOP strain
ε'_{EOP}	Strain Rate at the EOP Consolidation
$\rho; \rho_{EOP}$	Settlement; End of primary (EOP) settlement
σ'_p	Maximum past vertical effective stress
σ_{vf}	Final vertical effective stresses on the center of layer
σ_{vo}	Initial vertical total stresses on the center of layer
σ'_{vo}	Initial vertical effective stresses on the center of layer

REFERENCES

Alonso, U. R. and Aoki, N. (1988). "An alternative Solution for Deep Foundation in Baixada Santista". Anais do Simpósio Sobre Depósitos Quaternários das Baixadas Litorâneas Brasileiras, Rio de Janeiro, v. I: 5.11-5.26 (in Portuguese).

Andrade, M. E. S. (2009). Contribution to the Study of the Soft Clays of Baixada Santista. MSc Thesis, COPPE, UFRJ, 397p. (in Portuguese).

Aoki, N. and Angelino, C. N. (1994). "Deep Foundation in Baixada Santista". In: Solos do Litoral Paulista, ABMS - NRSP: 155-177 (in Portuguese).

Baguelin, F (1999). "La Détermination des Tassements Finaux de Consolidation: une Alternative à la Méthode d'Asaoka". Revue Française de Géotechnique n. 86: 9-17.

Falconi, F. F and Perez W. (2008). "Instrumented Static Load Test in a Steel Pile with Decreasing Cross Sectional Areas with Depth in Baixada Santista". IV CLBG e XI CNG, Coimbra-Portugal, Vol. 4, p. 147 a 154 (in Portuguese).

Golombek, S. (1965). "Towards an Orientation to the Design of Foundations in Baixada Santista" Instituto de Eng. de S. Paulo, Div. Téc. Estruturas, p. 30-39 (in Portuguese).

Gonçalves, H. L. S. and Oliveira, N. J. de. (2002a). "Comparison between Computed and Observed Settlements in a Santos' Building". 8o. Congresso Nacional de Geotecnia, Anais. v 2: 841-851, Abril de 2002 (in Portuguese).

Gonçalves, H. L. S. (2006). "Personnal Comunication".

Jamiolkowski, M.; Ladd, C. C. ; Germaine, J. T. and Lancellota, R. (1985). "New Developments in Field and Laboratory Testing of Soils". Proc. 11th ICSMFE, San Francisco, 1:57-153.

Kulhaway, F. H. and Mayne, P. N. (1990). "Manual on Estimating Soil Properties for Foundation Design". Report EL-6800, Electric Power Research Institute, Palo Alto.

Leroueil, S. (1996). "Compressibility of Clays: Fundamental and Practical Aspects". Journal Geotechnical Engineering, July: 534-543.

Maffei, C. E. M., Gonçalves, H. H. S., Pimenta, P. M. and Murakami, C. A. (2001). "The Plumbing of 2.2° Inclined Tall Building". 15th International Conference on Soil Mechanics and Geotechnical Engineering, Istanbul, vol. 3:1799-1802.

Machado, J. (1961). "Settlement of Structures in the City of Santos, Brazil". 5th ICSMFE, Paris. 1961. Proc. v. 1: 719-725. IPTPublication n. 629.

Massad, F. (1987). Sea-level Movements and the Preconsolidation of Brazilian Marine Clays In: INTERNATIONAL SYMPOSIUM ON GEOTECHNICAL ENG. OF SOFT SOILS, México City, v.1. p.91-98

Massad, F. (1988). "Load Transfer in Tubular Steel Piles, Installed in the Sediments of Baixada Santista". In: Simpósio sobre Depósitos Quaternários das Baixadas Litorâneas Brasileiras, Rio de Janeiro, ABMS, v.1. p.5.41-5.68 (in Portuguese).

Massad, F. (2004). "The use of Piezocones to Improve the Knowledge of the Marine Clays of Santos, Brasil". 9o CNG, Aveiro, Portugal, v.1, p. 309-318 (in Portuguese).

Massad, F. (2005). "Marine Soft Clays of Santos, Brazil: Building Settlements And Geological History". Proc. 16th ICSMGE. Osaka, Japan. v. 2:. 405-408.

Massad, F. (2009a). "Stress history and preconsolidation pressure evaluation from SCPTU". Proc. 17th ICSMGE. Alexandria, Egypt, v.2. p.961-964.

Massad, F. (2009b). "Behavior of Steel Piles in Baixada Santista: Instrumentation and Modeling". Engenharia de Fundações: passado recente e perspectivas. Homenagem ao Prof. Nelson Aoki. Suprema Gráfica e Editora Ltda, São Carlos, p. 103-116 (in Portuguese).

Massad, F. (2009c). "Marine Clays of Baixada Santista – Characteristics and Geotechnical Properties". São Paulo: Oficina de Textos, 247 p. (in Portuguese).

Olson, R. E. (1977). "Consolidation Under Time Dependent Loading". Journal of the Geotechnical Division, ASCE, 103 (GT1): 55-60.

Pacheco Silva, F (1953). "Controlling the Stability of a Foundation Through Neutral Pressure Measurements". International Conference on Soils Mechanics and Foundation Engineering, 3, Suisse, 1953, Proc... v. 1:299-301.

Rodrigues, J. C. (1965). "The Gelological Basis". In: The Baixada Santista – Geographical Aspects, Cap. 1o. Vol I: 23-48. Editora da USP (in Portuguese).

Suguio K. and Martin, L. (1981). "Progress in Research on Quaternary Sea Level Changes and Coastal Evolution in Brazil." Symposium on Variations in Sea Level in The Last 15,000 Years, Magnitude and Causes. Univ. South Caroline, USA.

Teixeira, A. H. (1960a). "Typical Subsoil Conditions and Settlement Problems in Santos, Brazil". Congreso Panamericano de Mecánica de Suelos e Ingenieria de Cimentaciones, México, v. I, p. 149-177 (in Portuguese).

Teixeira, A. H. (1960b). "Case History of Building Underlain by Unusual Conditions of Preconsolidated Clay (Santos, Brazil)". Congresso Panamericano de Mecánica de Suelos e Ingenieria de Cimentaciones, México, v. I, p. 201-215 (in Portuguese).

Teixeira, A. H. (1988). "Bearing Capacity of Pre-cast Concrete Piles in the Quaternary Sediments of Baixada Santista". Simp. Sobre Depósitos Quaternários das Baixadas Litor. Brasileiras, RJ, v. II: 5.1-5.25 (in Portuguese).

Teixeira, A. H. (1994). "Shallow Foudations in Baixada Santista". In: Solos do Litoral Paulista, ABMS - NRSP: 137-154 (in Portuguese).

The Embraport pilot embankment – primary and secondary consolidations of Santos soft clay with and without wick drains – Part 1

Rémy, J. P. P.*, Martins, I. S. M.**, Santa Maria, P. E. L.**, Aguiar, V. N.*,
Andrade, M. E. S.*
* Mecasolo Engenharia e Consultoria Ltda, Nova Friburgo, RJ, Brazil
** Rheology group – Soil Mechanics Division – COPPE –
Federal University of Rio de Janeiro, Brazil

KEYWORDS Soft clay, pilot embankment, secondary consolidation

ABSTRACT Routine *in situ* and laboratory investigations carried out for the basic design failed to provide reliable values for compressibility and consolidation parameters of soft clay layers at the Embraport site in Santos, Brazil. A pilot embankment was designed and built divided in three equal areas, two with wick drains in a square mesh at spacings of 1.2 m and 2.4 m respectively and one with no drain. Total primary and secondary compressions of the layer responsible for most of the settlement (the river lagoon sediments) have been computed from the compressibility parameters obtained from standard and special laboratory oedometer tests. It was expected to use an abacus relating strain and strain rate to the effective vertical stress built from the tests results to back-analyze the field measurements but this was found to be impossible.

1 INTRODUCTION

A multipurpose terminal is to be built at Barnabé Island on the Santos Channel opposite to Santos Harbour. Most of the area is to be used for containers storage and part of it will be reclaimed underwater.

The field investigations carried out for the basic design included 176 2¹/₂" percussion borings with SPT at every meter, 148 of them down to refusal, covering the whole area, including the quay and the vessels access channel areas, leading to an average distance between borings over the whole project area a little smaller than 100 m × 100 m, 220 "undisturbed" samples extracted from 4" and 6" borings, 375 vane-tests in 31 borings and 15 cone penetration tests with 78 pore pressure dissipation tests.

The subsoil is rather heterogeneous and consists of a layer of very soft to soft clayey alluvium (SPT from 0 to 4) with a total thickness varying between 22.0 m and 43.1 m. Within this layer, one or several sandy lenses or layers with SPT higher than 4 were found, mostly with individual thicknesses of 1 to 3 m, in 83 of the 148 deep borings. The sum of these layers and lenses thicknesses is up to 2 m in 34 borings, between 2 and 5 m in 34 borings and 5 to 11 m in 15 borings.

The basic design laboratory investigations consisted in 220 grain-size distribution curves, Atterberg limits, unit weight, natural water content and specific gravity and 141 standard stage loading oedometer tests, from which the conservative clay layer's parameters listed in Tables 1 and 2 were obtained to be used for the basic design of soft soil treatment, earthworks (fills and dikes) and quay, tanks and building foundations.

2 SOIL PARAMETERS CRITICALLY IMPORTANT FOR THE FINAL DESIGN AND CONSTRUCTION OF THE TERMINAL STILL NOT RELIABLY KNOWN AT THE END OF THE BASIC DESIGN

All parameters listed in Table 1 are fairly reliable and local variations from the design values do not imply in drastic impacts on the three key points of the design which are (a) settlements values, (b) time necessary for these settlements (and the associated undrained shear strength increase) to occur and (c) fills and dikes slopes safety factors. On the other hand, any variation of the compressibility, shear strength and consolidation parameters listed in Table 2 has a drastic impact on these key points, with special emphasis on the

TABLE 1 PHYSICAL PROPERTIES OF SOFT CLAY - BASIC DESIGN VALUES (*)

Depth	$\% < 5\,\mu$ (%)	w (%)	LL (%)	PI (%)	γ (kN/m³)	e_0
0 to 3 m						
3 to 6 m	36	73.5	78.2	47.1	15.6	2.08
6 to 9 m						
9 to 18 m	63	94.4	124.1	79.6	14.4	2.79
> 18 m	49	70.0	89.0	54.3	15.7	2.08

TABLE 2 COMPRESSIBILITY, SHEAR STRENGTH AND CONSOLIDATION PARAMETERS - BASIC DESIGN VALUES (**)

Depth	$C_c/(1+e_0)$	C_e/C_c	σ'_p (kPa)	S_u (kPa)	Su/σ'_p	c_h (cm²/s)
0 to 3 m			1.6 + 4.7z	0.5 + 1.2z		
3 to 6 m	0.28	0.112	14 + 4.7z	4.0 + 1.2z		
6 to 9 m			30 + 4.7z	9.0 + 1.2z	0.26	2.5×10^{-3}
9 to 18 m	0.36	0.126				
> 18 m	0.29	0.146	(z in meters)	(z in meters)		

(*) $\% < 5\mu$ = percentage of particles smaller than 0.005 mm, w = natural water content, LL = liquid limit, PI = plasticity index, γ = unit weight, e_0 = initial void ratio (after sampling), (**) C_c = compression index, C_e = expansion index, σ'_p = preconsolidation stress, S_u = undrained *in situ* shear strength, c_h = horizontal coefficient of consolidation

preconsolidation stress value, which is the soil parameter with most influence on each and everyone of the three key points listed above.

A real field value of preconsolidation stress greater than the design value will lead to overestimated settlements and understimated undrained shear strength of the soft clay. Hence, a safe design has to be based on a conservative design value of the preconsolidation stress chosen so that there will be low probability of the real field value to be smaller than the design value. However, the choice of a preconsolidation stress design value excessively conservative will lead to a very antieconomical design.

The scattering of the preconsolidation stress values obtained from the 141 oedometer tests carried out for the basic design is so high, as shown ahead in Fig. 4, that these values could not be used to define the preconsolidation stress design profile. This is a typical example of "highly scattered lab data resulting from poor quality samples and inappropriate testing such as ... doubling the load in oedometer test (ill defined σ'_p)" as pointed out by Ladd (2008). The only possible way of inferring the preconsolidation stress design profile was from the shear strength profiles given by the vane tests. These were used to obtain the design profile of σ'_p through the correlation between σ'_p and the *in situ* undrained shear strength Su, obtained from the vane test values Su (VT) by applying Bjerrum's (1972) correction factor to obtain the field values Su.

3 SOIL PROFILE PROPERTIES UNDER THE PILOT EMBANKMENT

The geotechnical investigations under the pilot embankment consisted of 5 percussion borings with SPT at every meter down to refusal, three 6" borings to obtain 4" undisturbed samples for laboratory tests, 4 CPT with 28 pore pressure dissipation tests and 67 vane tests in 5 borings, as shown in Fig. 1. The water content of SPT samples was measured in the material collected close to the bottom of every sample, which proved to be very helpful in defining the soil profile. The main aim of the pilot embankment and its geotechnical investigations was to provide the best information about the consolidation characteristics of the soft clay. The embankment's total area at elevation +3.00 m (which was the initial design elevation) is 150 m by 70 m. It is divided in 3 areas as shown in Fig. 1. Areas 1 and 2 are provided with 1.2 m × 1.2 m and 2.4 m × 2.4 m square meshes of wick drains penetrating 30 m from elevation +1.80 m down to elevation −28.20 m. Area 3 has no wick drain.

All undisturbed samples were extracted with stationary piston sampler under the supervision of the authors Martins and Aguiar present on the site to ensure that proper procedures were rigorously followed to provide the best possible samples. The 12 samples obtained in boring-SRA-203 under area 3 were very carefully sealed with paraffin, wrapped, packed and transported

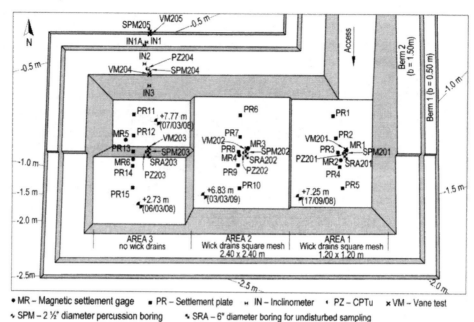

FIG. 1 *Pilot embankment – Location of soil investigations and of instruments for settlements and horizontal displacements measurements*

maintained in their proper upside vertical position to the soil mechanics laboratory of the rheology group at COPPE - UFRJ where they were tested. The test program consisted of unit weight, natural water content, specific gravity, grain size analysis, Atterberg limits and organic matter content for each sample. 42 standard and 32 special oedometer tests were performed under controlled temperature by Aguiar (2008) and Andrade (2009) on 70 mm diameter, 20 mm height specimens. The specimens were prepared according to Ladd and DeGroot (2003) recommendations. Fig. 2 shows the results of 7 oedometer tests carried out on undisturbed specimens and of one test carried out on a remoulded specimen, all from sample 6.

In respect to Coutinho's (2007) samples quality classification criteria all but three specimens from layer 6, were classified as excellent to fair.

Fig. 3 shows the soil profiles below the centers of areas 3, 2 and 1 with each layer description, the SPT values, the positions of the undisturbed samples and the positions of the settlement measuring magnets. In area 3, it also shows the initial vertical effective stress profile (σ'_v) and the preconsolidation stress profile (σ'_p) obtained from the oedometer tests performed on SRA-203 samples as discussed ahead.

As shown in Fig. 4, the preconsolidation stresses obtained at $\dot{\varepsilon}_v = 10^{-6}\,\text{s}^{-1}$ from the good quality samples tested at COPPE were much higher than the adopted values for the basic design, the same being true for the $C_c/(1+e_0)$

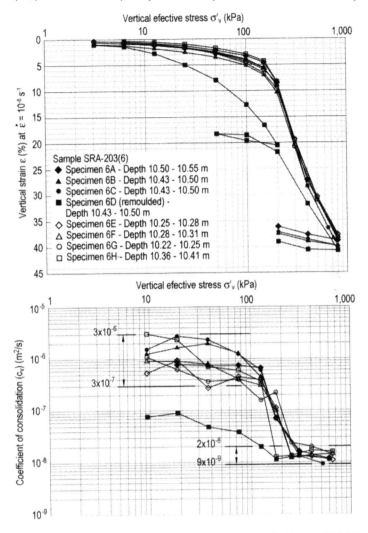

FIG. 2 *Results of standard oedometer tests on sample 6 of boring SRA-203*

values. Andrade (2009) observed that the σ'_p values at $\dot{\varepsilon}_v = 10^{-6}\,\text{s}^{-1}$ are, in average, 8% higher than the σ'_p values determined for 24h loading stages. The authors considered these COPPE new values to be very reliable since the same high repeatability of the results shown in Fig. 2 was obtained on all the clayey samples of SRA-203, and very good repeatability, although not as high, was obtained on the more sandy samples. The samples taken from borings SRA-201 and SRA-202 were tested in another laboratory and their results are not included in the present analysis. To define the σ'_p profile, a line passing through the highest value was chosen considering that these correspond to the highest quality specimens and that any degree of remoulding of the samples lower the value of σ'_p.

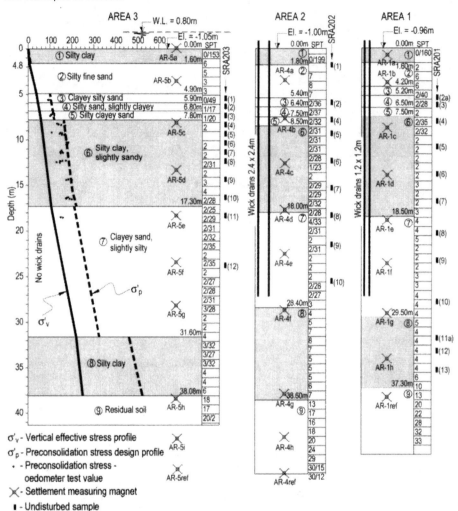

FIG. 3 *Soil profiles under areas 1, 2 and 3 with positions of undisturbed samples and settlement measuring magnets*

Table 3 shows the soils properties obtained from the soils investigations carried out in the foundation of the pilot embankment, including the physical properties of samples from SRA-201 and SRA-202.

4 PILOT EMBANKMENT CONSTRUCTION HISTORY AND INSTRUMENTATION

The pilot embankment construction was very carefully planned and carried out to avoid any local failure of the very soft clay layer down to 1.5 to 2 m depth. At first, nonwoven geotextile was laid on top of the very soft clay layer, covering the whole area. Then a 15 to 30 cm thick layer of quarry dust was hydraulically carefully spread into place. This layer was covered with a geogrid

TABLE 3 SOILS PROPERTIES FROM THE SOILS INVESTIGATIONS OF THE PILOT EMBANKMENT[*]

Layer (Fig. 3)	γ(kN/m^3)	$C_c/(1+e_0)$	C_r/C_c	c_{v1} (m^2/s)	c_{v2} (m^2/s)	c_{h1} (m^2/s)
1	13.0	0.36	0.12	—	—	—
2	20.0	—	—	—	—	1.5×10^{-4}
3	16.5	0.28	0.19	1.7×10^{-6}	6.0×10^{-7}	1.0×10^{-6}
4	17.6	0.16	0.27	1.4×10^{-6}	1.5×10^{-6}	8.3×10^{-6}
5	16.5	0.28	0.19	1.7×10^{-6}	6.0×10^{-7}	1.0×10^{-6}
6	15.0	0.56	0.11	6.5×10^{-7}	1.5×10^{-8}	4.9×10^{-7}
7	18.1	0.16	0.27	1.5×10^{-6}	1.3×10^{-6}	2.1×10^{-5}
8	14.6	0.56	0.11	6.5×10^{-7}	1.5×10^{-8}	4.9×10^{-7}

(*) c_{v1} and c_{h1} are for $\sigma'_v \leq \sigma'_p$, and c_{v2} is for $\sigma'_v > \sigma'_p$ (c_{v1} and c_{v2} from oedometer tests, c_{h1} from CPT porepressure dissipation tests)

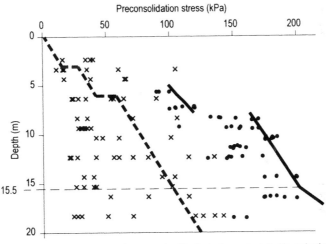

× Results from 24h stages oedometer tests of basic design geotechnical investigations (see Fig. 2)

Preconsolidation stress profile for basic design

• σ'_p at $\dot{\varepsilon}$ = 10^{-6} s^{-1} from oedometer tests run on undisturbed samples of boring SRA203 (see Fig. 5)

Preconsolidation stress profile from tests on samples of boring SRA203 (see Fig. 5)

FIG. 4 *Preconsolidation stress values and profiles*

with a strength of 800 kN/m × 80 kN/m and the fill material was slowly put into place, first being dumped under the water level and then as soon as access was possible at low tide, poured by trucks and bulldozers. When elevation +1.80 m was reached, the wick drains were driven requiring that holes be first drilled through the embankment and the geogrid. The embankment was then completed up to elevation +3.00 m.

After obtaining the first COPPE oedometer tests high preconsolidation stress values, it was decided to heighten the pilot embankment in order to apply a vertical stress higher than the preconsolidation stress. The highest embankment elevations in each area and the dates when they were reached as well as the locations of the instruments installed to monitor settlements and horizontal displacements are shown in Fig. 1. When the decision to increase the embankment thickness was taken, the north half of area 3 had already been heightened to elevation +7.77 m, immediately after the embankment was completed to design elevation +3.00 m, with the aim to provoke a foundation failure towards inclinometers IN1, IN2 and IN3. While the heightening of the fill was taking place, the stresses in the geogrid in the north south direction were measured with the installed strain gauges. The embankment did not show apparent deformations, almost no stress change was recorded in the strain gauges and very little lateral displacements were registered in the inclinometers. The analysis of this behaviour is beyond the scope of the present work.

The pilot embankment was also instrumented with full profile hydraulic settlement gauges, electrical and open pipe piezometers, but these instruments, the inclinometers and the strain gauges of the geogrid were read for a too short period of time to provide helpful data.

5   SPECIAL LABORATORY TESTS

Secondary consolidation was observed after unloading in 26 long term oedometer tests. Stress relaxation tests where the vertical displacements were restrained and the vertical stress measured were also carried out.

Four specimens trimmed in sample SRA-203(8), were consolidated under 800 kPa following the same procedure, unloaded to different stress/OCR values and the strains were then measured for 20 days under constant stress and constant temperature (20°C ± 1°). The specimen unloaded to the OCR value of 1.60 stopped expanding and started to compress whereas the specimens unloaded to OCR values of 2.00, 2.29 and 2.67 leveled off. From these and all other results from long term oedometer tests performed on other samples, Andrade (2009), following the procedure suggested by Feijó and Martins (1993), established that the end of secondary compression line could be drawn as being the line equivalent to an OCR value of 2.1 relative to the $\dot{\varepsilon} = 10^{-6} s^{-1}$ line. In their analysis the authors did not discriminate between the $\dot{\varepsilon} = 10^{-6} s^{-1}$ line and the end of primary compression line due to the proximity of these two lines observed in all oedometer tests results. This value can be compared to the equivalent OCR value close to 2.0 relative to E.O.P. found by Feijó (1991) for the end of secondary compression of Sarapuí clay in Rio de Janeiro, and to 1.7 relative to 24 hour stages published by Martins et al. (2009) for various Rio de Janeiro soft clays.

The preconsolidation stresses σ'_p in layers 3 to 6 were established from the oedometer tests results as seen on Fig. 4. It can be seen that, in layer 6, above 15.5 m depth the lab results show σ'_p values higher than 2.1 σ'_v. This is attributed to dune effect (Massad, 1989). The dune effect was considered in layer 7 and the secondary compression preconsolidation effect was considered in layers 1 and 8 adopting $\sigma'_p = 2.1\sigma'_v$.

6   PILOT EMBANKMENT MONITORING DATA

Fig. 5 shows the course of the elevation of the embankment and of the total settlements in the center of area 1 measured through settlement plate PR3, in the center of area 2 measured through settlement plate PR8 and close to the center in area 3 measured through settlement magnet AR-5a of gauge MR5 for lack of readings of damaged settlement plate PR13. Due to the lack of readings of the other instruments, the monitoring data analyzed to assess the behaviour of the soft marine clay under the pilot embankment is limited to the settlement data provided by the settlement plates and the magnetic settlement gauges.

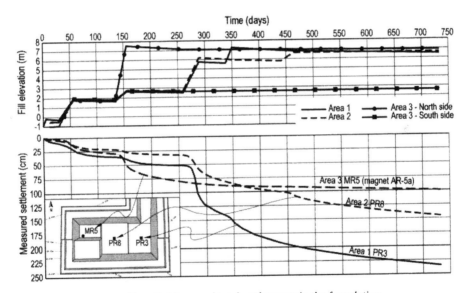

FIG. 5 *Measured total settlements in the foundation*

The settlements and settlement rates measured in early October 2009, 2 years after the beginning of construction, are shown in Table 4.

Fig. 6 shows the evolution with time of the elevation of the top of the embankment and of layer 6 compression in the three areas centers.

Table 5 shows the values of measured layers compressions, compression rates, calculated strains and strain rates in October 2009.

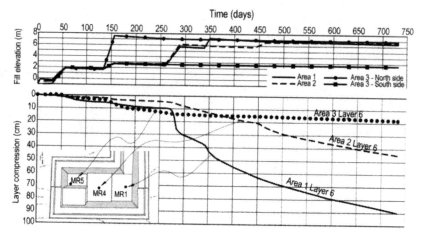

FIG. 6 *Measured compression of layer 6*

TABLE 4 FOUNDATION SETTLEMENTS AND SETTLEMENT RATES IN EARLY OCTOBER 2009

Area	Settlement	Strain	Settlement rate	Strain rate
1	236 cm	Not applicable	3.2 cm per month	Not applicable
2	146 cm	Not applicable	2.9 cm per month	Not applicable
3	101 cm	Not applicable	0.8 cm per month	Not applicable

TABLE 5 VALUES OF MEASURED LAYERS COMPRESSIONS AND COMPRESSION RATES AND OF CALCU-LATED STRAINS AND STRAIN RATES IN EARLY OCTOBER 2009

Layers	Compression (cm) / Strain (%)		
	Area 1	Area 2	Area 3
1, 2, 3, 4, 5[(*)]	80.8/10.8	50.9/6.0	47.6/6.1
6	90.0/8.2	45.0/4.7	18.0/1.9
7, 8[(**)]	65.1/3.5	50.5/2.5	35.0/1.7

Layers	Compression rate (cm/month) / Strain rate (s^{-1})		
	Area 1	Area 2	Area 3
1, 2, 3, 4, 5[(*)]	$0.1/5.8 \times 10^{-11}$	$0.1/4.8 \times 10^{-11}$	$0.3/1.6 \times 10^{-10}$
6	$2.5/9.5 \times 10^{-10}$	$1.7/7.2 \times 10^{-10}$	$0.3/1.2 \times 10^{-10}$
7, 8[(**)]	$0.6/1.3 \times 10^{-10}$	$1.1/2.1 \times 10^{-10}$	$0.2/3.8 \times 10^{-11}$

(*) Most of the compression still in process is believed to occur predominantly in layer 1, (**) most of the compression still in process is believed to occur predominantly in layer 8

7 PRIMARY COMPRESSION AND SECONDARY COMPRESSION

To analyze the settlement data presented in Figs. 5 and 6, the vertical stress increase and the final vertical effective stress (compared to the preconsolidation stress) have to be known. Fig. 7 shows the profiles of final vertical

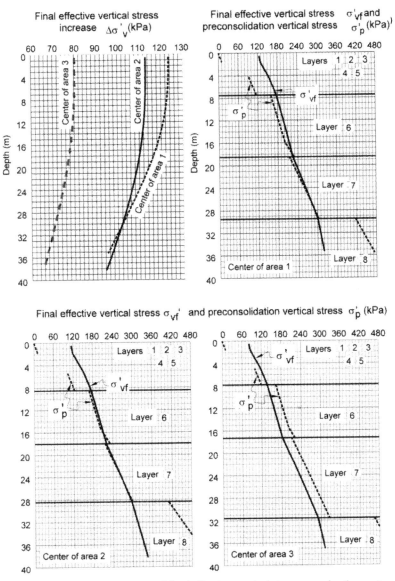

FIG. 7 *Vertical stress increases and final effective vertical stresses under the centers of areas 1, 2 and 3*

stress increases calculated under the centers of areas 1, 2 and 3. The stress increase calculations were done through Holl's (1940) formulas to compute $\Delta\sigma$ at depth z under the corner of loaded rectangles. As can be seen, the stress increases are different under every area. Fig. 7 also shows the final effective vertical stresses under the centers of the three areas compared to the profiles of preconsolidation stresses. The term "final" used to characterize both vertical stresses and vertical stress increases means the values calculated for the final situation reached after full completion and stabilization of both primary and secondary consolidation settlements, taking into consideration the decrease of effective stresses caused by the progressive partial submersion of the embankment as settlements develop. In order to obtain these "final" values, calculations had to be done by successive iterations converging step by step.

In area 1, the final effective vertical stresses are clearly higher than the preconsolidation stresses σ'_p in layers 1 to 5, a little higher than σ'_p in layers 6 and 7 and much smaller than σ'_p in layer 8. In area 2, the final effective vertical stresses are clearly higher than σ'_p in layers 1 to 5, about equal to σ'_p in layers 6 and 7 and much smaller than σ'_p in layer 8. In the center of area 3, the final effective vertical stresses are higher than σ'_p in layers 1 to 5 and smaller than σ'_p in layers 6, 7 and 8.

Fig. 8 illustrates the way primary and secondary compressions were computed, in the case when the final effective vertical stress is higher than the preconsolidation stress on the left and in the case when the final effective vertical stress is lower than the preconsolidation stress on the right. Primary compression was computed through the common well known formulas using C_r and C_c in the recompression and virgin compression ranges respectively, and secondary compression was computed through the following formulas for 1 m thick sublayers.

$$\Delta H_S = H \frac{C_c}{1 + e_0} \left(1 - {}^{C_e}\!/_{C_c}\right) \log (2.1) \quad \text{and} \quad \Delta H_S = H \frac{C_c}{1 + e_0} \left(1 - {}^{C_e}\!/_{C_c}\right) \log \left(2.1 \frac{\sigma'_{vf}}{\sigma'_p}\right)$$

when $\sigma'_{vf} > \sigma'_p$ and $\sigma'_{vf} < \sigma'_p$, respectively.

The preconsolidation stresses and calculated final effective vertical stresses at mid-height of each layer and the primary and secondary compressions of layers 1 to 8, considering the soil between the drains in areas 1 and 2 to be perfectly undisturbed are shown in Tables 6, 7 and 8.

It is the authors' understanding that any remoulded zone around the drains would lead to higher settlements as clearly illustrated by Fig. 2. The discussion of this subject is beyond the scope of this work.

8 METHODOLOGY EXPECTED TO BE APPLIED TO ANALYZE THE PILOT EMBANKMENT SPECIFIC LAYERS COMPRESSION DATA

A schematic path during field preloading was proposed by Bjerrum (1972). A schematic path during a multiple-stage loading oedometer test was proposed by Leroueil et al. (1985). Both propositions are reproduced from the original papers in Fig. 9. On the left side of Fig. 9, the authors added the values of the "Rate of secondary consolidation" ($\dot{\varepsilon}$) expressed in s^{-1} calculated from the values per year indicated in Bjerrum's figure. The right side of Fig. 9, shows the figure of the paper by Leroueil et al. (1985).

TABLE 6 PRECONSOLIDATION STRESSES AND CALCULATED FINAL EFFECTIVE VERTICAL STRESSES AT MID HEIGHT OF EACH LAYER AND PRIMARY AND SECONDARY COMPRESSIONS IN AREA 1

	Layer 1	Layer 2	Layer 3	Layer 4	Layer 5	Layer 6	Layer 7	Layer 8	Total
σ'_{vf}	125.9	141.3	157.4	165.4	173.4	201.6	263.1	310.5	
σ'_p	5.0		95.7	103.8	112.1	187.1	259.2	446.7	
ΔH_p	0.67			0.22[*]		0.43	0.23	0.06	1.61
ΔH_s	0.16			0.15[*]		1.77	0.00	0.51	2.59
ΔH_t	0.83			0.37[*]		2.20	0.23	0.57	4.20
$\Delta H_m/\Delta H_t$	65%			71%[*]		41%		82%[**]	56%

TABLE 7 PRECONSOLIDATION STRESSES AND CALCULATED FINAL EFFECTIVE VERTICAL STRESSES AT MID HEIGHT OF EACH LAYER AND PRIMARY AND SECONDARY COMPRESSIONS IN AREA 2

	Layer 1	Layer 2	Layer 3	Layer 4	Layer 5	Layer 6	Layer 7	Layer 8	Total
σ'_{vf}	115.8	136.5	157.6	165.0	172.3	198.3	259.5	322.4	
σ'_p	5.7		106.3	113.4	121.2	192.5	258.4	470.8	
ΔH_p	0.76			0.17[*]		0.26	0.21	0.06	1.46
ΔH_s	0.18			0.15[*]		1.52	0.00	0.58	2.43
ΔH_t	0.94			0.32[*]		1.78	0.21	0.64	3.89
$\Delta H_m/\Delta H_t$	31%			66%[*]		25%		60%[**]	38%

TABLE 8 PRECONSOLIDATION STRESSES AND CALCULATED FINAL EFFECTIVE VERTICAL STRESSES AT MID HEIGHT OF EACH LAYER AND PRIMARY AND SECONDARY COMPRESSIONS IN AREA 3

	Layer 1	Layer 2	Layer 3	Layer 4	Layer 5	Layer 6	Layer 7	Layer 8	Total
σ'_{vf}	82.8	101.7	121.3	127.9	134.5	160.6	237.4	301.6	
σ'_p	5.0		102.7	109.3	116.0	187.3	268.8	492.2	
ΔH_p	0.61			0.10[*]		0.18	0.15	0.04	1.08
ΔH_s	0.16			0.15[*]		1.20	0.00	0.30	1.81
ΔH_t	0.77			0.25[*]		1.38	0.15	0.34	2.89
$\Delta H_m/\Delta H_t$	36%			77%[*]		13%		72%[**]	35%

[*] layers 3, 4 and 5 altogether, [**] layers 7 and 8 altogether.

where: σ'_{vf} = final vertical effective stress at mid-height of whole layer (kPa),
σ'_p = preconsolidation stress at mid-height of whole layer (kPa), ΔH_p = layer primary compression (m), ΔH_s = layer secondary compression (m), ΔH_t = layer total compression (m), ΔH_m = measured layer compression (m) in October 2009.

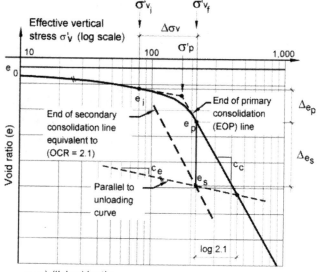

e_i = initial void ratio

$\sigma'v_i$ = initial vertical effective stress

$\Delta\sigma_V$ = vertical stress increase

$\sigma'v_f$ = final vertical effective stress

e_p = final void ratio at end of primary consolidation
(without secondary)

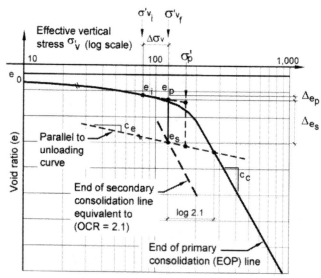

e_s = final void ratio at end of primary and secondary
consolidation

$\sigma'p$ = preconsolidation stress

Δe_p = variation of void ratio corresponding to primary
compression

Δe_s = variation of void ratio corresponding to secondary
compression

FIG. 8 *Primary compression and secondary compression*

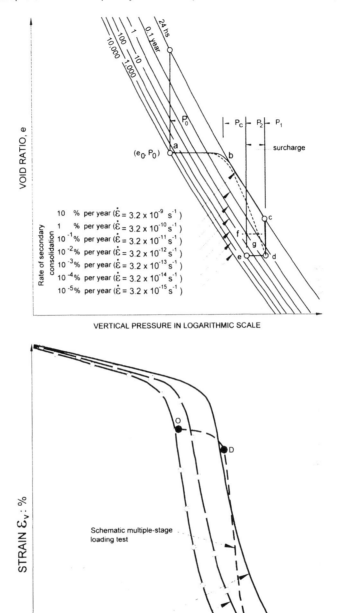

FIG. 9 *Schematic path proposed by Bjerrum (1972) and Leroueil et al. (1985)*

Leroueil et al. (1985) stated:

> Taylor and Merchant (1940) were the first to suggest a model in which the rate of change in void ratio is a function of the effective stress, the void ratio and the rate of change in effective stress. This suggestion has been followed by numerous researchers... The rheological models proposed were seldom assessed experimentally or only on the basis of a few laboratory test results. Experimental studies, however, have been performed on natural clays... and on resedimented clays... but in each study only one type of test was used... It is thus difficult to obtain an overall view of the rheological behaviour of clays from these studies. As a result this abundant literature has modified neither the common practice based on the Terzaghi theory nor the way of thinking on clay behaviour.

In the authors' knowledge not much has changed from 1985 to 2010, and for sure, practicing engineers continue not disposing of any practical tool to predict or to back-analyze the simultaneous evolution of primary and secondary settlements in soft clayey soils under embankments. For the last 30 years or so, the main trend of thought of the most prolific Brazilian researchers in the study of the behaviour of marine Santos soft clay has sustained that secondary consolidation only starts to take place at the end of primary consolidation. This working hypothesis has been invoked to justify the fact that all settlements measured during the usual period of monitoring of test embankments, a year more or less in most cases, can be considered to be entirely primary settlements. The measured settlement curves, mainly the settlement plates curves, can then be analyzed using Asaoka's (1978) method based on the consideration that for $\overline{U} \geq 33\%$ the series solution of Terzaghi's theory could be replaced by its first term which is $\overline{U} = 1 - 0.811\,e^{\frac{-\pi^2}{4}T}$. It is then justifiable to use Asaoka's construction to obtain the total primary settlement and the consolidation coefficient, assuming that after a few months the average consolidation ratio under field test embankments could be considered to be higher than 33%. In the case of radial drainage considering equal strain theory, the average consolidation ratio is given by an expression presented by Barron (1948) similar to the one above mentioned, so that Asaoka's construction can also be applied. In both cases, the consolidation coefficient and total primary settlement thus obtained have been considered sound and trustworthy. This approach has led to the endlessly repeated conclusion that the back-analyzed field values of the consolidation coefficient are usually about 50 times higher than the laboratory consolidation coefficient and that this fact is totally explained and justified by the existence of providential very thin, obviously continuous, layers of pervious sand. It is worth mentioning that no lense of pervious sand was present in any of the layer 6 samples.

The authors consider, first, this explanation too much far-fetched to be acceptable at once and, second, that some basic points could not, and should not, have been overviewed for so many years, as has been the case. The first point which has been systematically overviewed is the fact that the consolidation coefficient values are very different in the range of stresses up to the preconsolidation stress and in the range of stresses beyond the preconsolidation stress. This was very clearly illustrated in Fig. 2 and Bjerrum (1972), just as an example, can be quoted in this respect:

> The rate of pore pressure dissipation can in principle, be computed on a similar basis as the consolidation settlement, i. e. taking into account that the consolidation properties are different in the two ranges, the first one representing the increase in effective stresses in the range from p_o to p_c and the second one representing the increase in effective stress beyond p_c.

This point was clearly brought up by Gonçalves (1992):

> It is well known that the c_v value of a preconsolidated soil can be 10 to 100 times higher than the c_v value of a normally consolidated soil. Some papers which compare the behaviours of the soils from Santos lowlands in the field and in the laboratory do not state how the c_v value was obtained and others admit to have used the c_v value corresponding to stresses $\sigma_v' > \sigma_p'$. If in these analysis the c_v value for the preconsolidated range had been used, the differences between field and laboratory behaviours would probably have been smaller.

The second point which has also been systematically overviewed is the fact that much evidence has been brought up in the last decades in the technical literature showing that secondary consolidation under embankments in the field occurs simultaneously with primary consolidation. The argument repeatedly used by the researchers to justify the working hypothesis of secondary consolidation only starting to occur after the end of primary consolidation is based on the fact that since long term settlements measured in the area of Santos vary rather linearly with the logarithm of time, then the formulation "$\Delta e = C\alpha \log \Delta t$" holds true and since $C\alpha$ is historically considered to express the evolution of secondary settlements after primary settlements have ended, then the straight line obtained by plotting settlements versus log of time has been taken as being the proof that the working hypothesis is true. Since to the authors it is clear that long field term settlements will vary, for a long period before coming close to stabilization, rather linearly with the logarithm of time, whether they are only primary, or only secondary or the sum of both, the above argument put forward by the researchers is no more than a sophism.

Based on the diagrams in Fig. 9, it was the authors' expectation that they would be able to produce an abacus, specific for the soft clay of layer 6 tested at COPPE, from the laboratory test results. The strain (ε_t) and strain rate ($\dot{\varepsilon}_t$) at time t in layer 6 calculated from the layer compression measurements under the pilot embankment when plotted together on this abacus would immediately yield the mean effective stress and mean primary consolidation ratio of the layer at time t. The left side of Fig. 10 illustrates that the strain at time $t(\varepsilon_t)$ would be entered in ordinate as a horizontal line extended to the right until it meets the strain rate line representing the strain rate $(\dot{\varepsilon}_t)$ occurring in the layer at time t, at point A. From point A, the value of σ'_{vt} = mean effective stress in the layer at time t would be directly read in the abacus giving the mean primary consolidation ratio (U_t) at time t calculated as $U_t = (\sigma'_{vt} - \sigma'_{vi})/(\sigma'_{vf} - \sigma'_{vi})$.

After removing the fill surcharge or after stopping applying vacuum in the drains, this resulting in a vertical unloading effective vertical stress change equal to $-\Delta\sigma'_{v1}$, the situation would correspond to point B and after terminal operation reaches its normal operation capacity with the application of an effective vertical stress increase $\Delta\sigma'_{v2}$, the situation would be illustrated by point C.

The special oedometer tests mentioned in item 5 aimed to provide the necessary information to produce the abacus representative of the behaviour of layer 6. The emphasis was put on layer 6 for the following reasons. From the soil profiles shown in Fig. 3, it was expected that most of the foundation compression would occur in layers 1 and 6, which Massad (1989) respectively classifies as "Mangrove clay" and "SFL", which stands for "Sedimentos flúvio-lagunares", that is "river-lagoon sediments". No undisturbed sample could be recovered from layer 1 and the compressibility and consolidation properties of layer 1 clay could be obtained from the monitoring data. So all emphasis for the undertaken studies based on laboratory tests was put on layer 6. The right side of Fig. 10 shows the abacus built combining the results of all special oedometer tests which provided strain rates determination during secondary consolidation with the compressibility curve of sample SRA-203(4).

As soon as the abacus was ready, the authors perceived at once that it could not be representative of the behaviour of layer 6 in the field under area 3 where strain rates at the beginning of consolidation would be less than $10^{-10} \, \text{s}^{-1}$ and would plot outside the abacus. This indicates that the abacus of Fig. 10 is only representative of the behaviour of a 2 cm thick sample of layer 6 soft clay and that for any given clay, there exists a different specific abacus which represents the behaviour of the clay, for each layer thickness. The question which then arose is "can the abacus for layer 6 soft clay field layer be obtained through some realistic model from the laboratory sample abacus of Fig. 10?" This question is treated in the 2nd part of this paper.

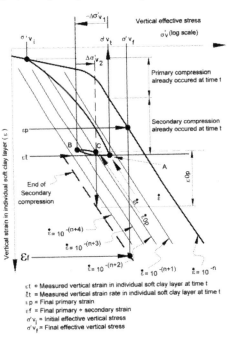

εt = Measured vertical strain in individual soft clay layer at time t
ε̇t = Measured vertical strain rate in individual soft clay layer at time t
εp = Final primary strain
εf = Final primary + secondary strain
σ'vi = Initial effective vertical stress
σ'vf = Final effective vertical stress

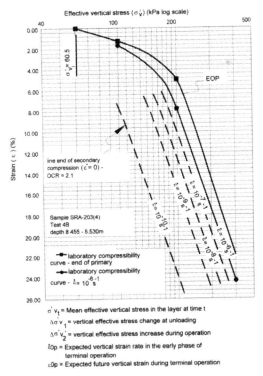

σ'vt = Mean effective vertical stress in the layer at time t
Δσ'v1 = vertical effective stress change at unloading
Δσ'v2 = vertical effective stress increase during operation
ε̇0p = Expected vertical strain rate in the early phase of terminal operation
ε0p = Expected future vertical strain during terminal operation

FIG. 10 *Methodology expected to be used to analyze layer 6 compression under the pilot embankment and abacus built through oedometer tests results*

References

Aguiar, V. N. (2008). *Características de adensamento da argila do canal do Porto de Santos na região da Ilha Barnabé* M.Sc. Thesis, COPPE/UFRJ, Rio de Janeiro, RJ, Brazil. (in Portuguese).

Andrade, M. E. S. (2009). *Contribuição ao estudo das argilas moles da cidade de Santos* M.Sc. Thesis, COPPE/UFRJ, Rio de Janeiro, RJ, Brazil. (in Portuguese).

Asaoka, A. (1978). *Observational procedure of settlement prediction* Soils and foundations, Japanese Society of Soil Mechanics Foundation Engineering, v.18, n. 4, pp. 87-101.

Barron, R. A. (1948). Consolidation of fine-grained soils by drain wells, *Transactions*, ASCE, v. 113, p. 718-742.

Bjerrum, L. (1972). Embankments on soft ground. Performance of earth and earth supported structures, ASCE, pp. 1-54.

Coutinho, R. Q. (2007) *Characterization and Engineering Properties of Recife Soft Clays - Brazil*, Characterization and Engineering Properties of Natural Soils, Taylor and Francis - Balkema, Editors Tan, Phoon, Hight and Leroueil, vol. 3, pp. 2049-2100.

Feijó, R. L. (1991). *Relação entre a compressão secundária, razão de sobreadensamento e coeficiente de empuxo no repouso*, M.Sc. thesis, COPPE/UFRJ, Rio de Janeiro, RJ, Brazil (in Portuguese).

Feijó, R. L. and Martins, I. S. M. (1993). *Relação entre a compressão secundária, OCR e K_0*. COPPEGEO 93, In: Simpósio Geotécnico Comemorativo dos 30 anos da COPPE/UFRJ, Rio de Janeiro, RJ, Brazil, pp. 27-40 (in Portuguese).

Gonçalves, H. H. S. (1992). *Análise crítica através de modelos visco-elásticos dos resultados de ensaios de laboratório para previsão de recalques dos solos da baixada santista*. PhD thesis, USP, São Paulo, Brazil (in Portuguese).

Holl, D. L. (1940). Stress transmission in earths, proc. High res. Board, vol. 20, pp. 709-721.

Ladd, C. C. and De Groot, D. J. (2003). Recommended practice for soft ground characterization, Arthur Casagrande Lecture, 12th Panamerican Soil Mechanics Conference.

Ladd, C. C. (2008). *Commentary: Soft ground geotechnics*, geo-strata, march/april, pp. 10-11.

Leroueil, S.; Kabbaj, M.; Tavenas, F. and Bouchard, R. (1985). *Stress-strain rate relation for the compressibility of sensitive natural clays*, Geotechnique, vol. 35, n° 2, pp. 159-180.

Martins, I. S. M.; Santa Maria, P. E. L. and Santa Maria, F. C. M. (2009) *Laboratory behaviour of Rio de Janeiro soft clays. Part 1: index and compression properties*, Discussion, soils and rocks, v. 32:2, pp. 100-103, São Paulo.

Massad, F. (1989). *Settlements of earthworks construction on Brazilian marine soft clays in the light of their geological history.* In: XII International Conference on Soils Mechanics and Foundation Engineering, Rio de Janeiro. Proceedings. v. 3. pp. 1749-52.

Taylor, D. W. and Merchant, W. (1940). *A theory of clay consolidation accounting for secondary compression,* Journal of Mathematics and Physics, vol.19, n.3, pp. 167-185.

The Embraport pilot embankment – primary and secondary consolidations of Santos soft clay with and without wick drains – Part 2

Rémy, J. P. P.*, Martins, I. S. M.**, Santa Maria, P. E. L.**, Aguiar, V. N.*,
Andrade, M. E. S.*
* Mecasolo Engenharia e Consultoria Ltda, Nova Friburgo, RJ, Brazil
** Rheology group – Soil Mechanics Division – COPPE –
Federal University of Rio de Janeiro, Brazil

KEYWORDS Soft clay, pilot embankment, secondary consolidation

ABSTRACT Taylor and Merchant's formulation has been used to back-analyze the behaviour of the area of the pilot embankment without drains and a hybrid method called "primary Barron + secondary pseudo Taylor and Merchant" has been used for the two areas with wickdrains. The coefficient of consolidation values back-analyzed through these methods decrease as the consolidation ratio increases and their order of magnitude is in remarkably good agreement with the laboratory values. It is shown that the use of Asaoka's method leads to values of final primary settlement, and coefficient of consolidation which can be much in error only due to the fact that the simultaneous occurrence of primary and secondary consolidations is not taken into account.

1 INTRODUCTION

This paper is the continuation of *"The Embraport pilot embankment – primary and secondary consolidations of Santos soft clay with and without wick drains – Part 1"* (Rémy et al., 2010).

2 TENTATIVE MODEL TO ANALYZE THE BEHAVIOUR OF THE THICK CLAY LAYERS

As mentioned by Leroueil et al. (1985), many formulations have been put forward to try and model secondary consolidation occurring together with primary consolidation. Taylor and Merchant's (1940) theory (theory A) was the first one published in the technical literature stating a basic postulate's to express the evolution of the soil void ratio with time during secondary consolidation. This postulate states "that the speed of occurrence of secondary compression is proportional to the undeveloped secondary compression" and introduces the soil parameter (μ) called the "coefficient of secondary compression". The mathematical solution to compute the "aggregate consolidation ratio" based on the above postulate was given by Taylor and Merchant showing that the evolution of this ratio depends on two parameters. The first one (r) is the ratio of the primary settlement over the total settlement, based on the understanding that secondary compression is finite and can be estimated. The second one is a dimensionless term (F) called the "Secondary compression factor" such as:

$$F = \frac{\mu t}{rT} = \frac{\mu H_d^2}{rc_v}$$

The original paper includes a graph showing the aggregate consolidation ratio (referred in this paper as U_{p+s}) where it can be appreciated that this is the ratio of the strain at time t over the total (primary plus secondary) strain plotted as a function of time factor T for the case of unidimensional vertical compression with only vertical drainage for an r value of 0.70 and for various values of the F factor.

In Taylor and Merchant's own words, this formulation is "based largely on physical intuition" and

> the main value of this theory is not in the expression of a secondary compression time law, but in the rational explanation of the superimposed action of the two very different physical laws of time rate, allowing the prediction of the action in thick strata from the action in laboratory tests.

This "prediction" would be the answer to the authors' previously posed question (Rémy et al., 2010) "can the abacus for layer 6 soft clay field layer be obtained through some realistic model from the laboratory sample abacus" and if so is this usable and reliable?

Christie (1964) showed that the formulas published by Taylor and Merchant are mistaken, but that their curves are correct, in accordance with the right formulas reestablished by Christie (1964) in terms of excess pore pressure. Furthermore, Christie showed that Gibson and Lo's (1961) formulation is identical to Taylor and Merchant's although expressed in different terms. The mathematical formulation for Taylor and Merchant's theory was obtained by Carvalho (1997) in terms of void ratio. This was used by the authors to compute the aggregate consolidation ratio for r = 0.70 and F values of 0, 0.1 and 10 obtaining perfect agreement with Taylor and Merchant's published curves. It was also used for all further calculations of primary plus secondary compressions included ahead in this paper.

A number of long term oedometer tests has been performed in the laboratory of the Rheology group of the Soil Mechanics Division at COPPE, Rio de Janeiro. Fig. 1 shows the measured curves and the theoretical curves obtained from Taylor and Merchant's theory which best fit the experimental curves for one clay sample of the Rio de Janeiro area, Senac clay (Martins, 2005), and

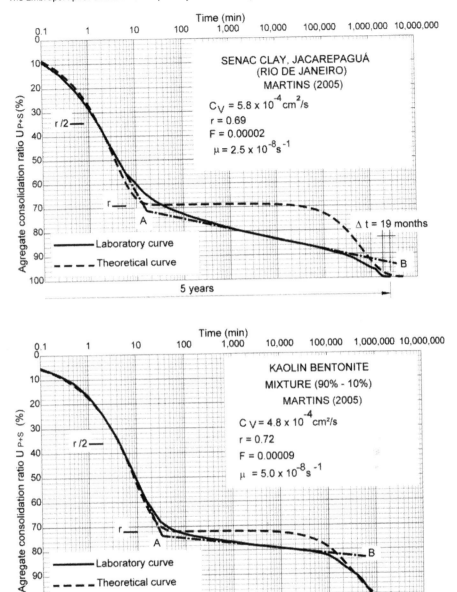

FIG. 1 *Comparison of long term oedometer tests in Senac clay and kaolin-bentonite mixture with Taylor and Merchant's theory*

for a specimen trimmed from a sample prepared through a mixture of kaolin (90%) and bentonite (10%) (Martins, 2005).

Line AB shows for each test the line from which the value of the coefficient of secondary compression $C\alpha$ would be normally determined from the oedometer test result. As can be seen this line only fits the test results up to around 100,000 minutes, that is about 70 days, after that a much higher value of Cα would be required to fit the test results.

The best fit is obtained by varying the three parameters r, c_v and μ and Fig. 1 shows the values which provided the best fits. These values and the values obtained for Sarapuí clay (Vieira, 1988), are compared in Table 1.

TABLE 1 VALUES OF R, C_v AND μ FOR SENAC CLAY, SARAPUÍ CLAY AND KAOLIN-BENTONITE MIXTURE

	Senac clay	Sarapuí clay	Kaolin+bentonite
r	0.69	0.79	0.72
c_v $(10^{-8}m^2/s)$	5.8	1.2	4.8
μ $(10^{-7}s^{-1}/s)$	0.25	1.5	0.5

The test on the Senac clay lasted for 5 years and it can be seen that the compression totally stabilized after 3.4 years, confirming that the secondary compression is finite and ends at some time. As can be seen on Fig. 1 Taylor and Merchant's theory agrees rather well with the experimental curves at the beginning and at the end of the consolidation process but fails to provide good agreement in between, where the measured secondary compression is consistently higher than the theoretically predicted one, which means, in other words, that the secondary consolidation on 2 cm thick samples proceeds faster than modeled by Taylor and Merchant's theory.

Fig. 2 shows Taylor and Merchant's aggregate consolidation ratio versus time for samples or layers, double drained (H_d = drainage distance = half the thickness) with thicknesses varying from 20 mm (lab sample), up to 20 m for a clay with $c_v = 1 \times 10^{-8}$ m^2/s, $\mu = 1.5 \times 10^{-7}$ s^{-1} and with r = 0.30.

Line AB drawn on Fig. 2 for H_d = 10 mm, that is for the standard oedometer test, is the line from which the value of the coefficient of secondary compression Cα would be normally determined from the laboratory consolidation curve. According to Taylor and Merchant's formulation, it can be seen that the inclinations of the lines drawn in Fig. 2 in the range Uprimary > 90% would not be constant through time for H_d up to 40 mm, as already observed on the actual laboratory tests curves shown in Fig. 2. It can also be seen that the inclinations of the curves are about the same for H_d > 100 mm and it is clear that these inclinations are totally different from the laboratory Cα line.

Table 2 shows the time to reach U_{p+s} equal to 99.9% from Taylor and Merchant's theory in a clay with, $\mu = 1.5 \times 10^{-7}$ s^{-1} and r = 0.3, for c_v equal to 4.0×10^{-7} m^2/s and to 1.0×10^{-8} m^2/s.

It seems reasonable to expect that Taylor and Merchant's theoretical curves might be rather realistic at the beginning and at the end of the consolidation process and might underestimate the settlement (or strain) in the middle of the process as was observed on the 2 cm thick laboratory samples (see

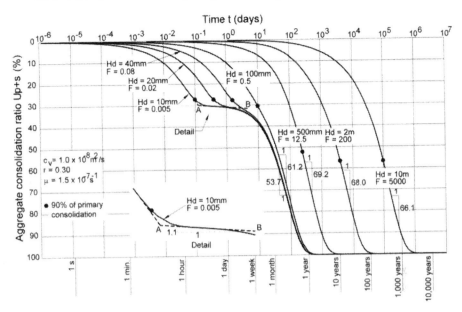

FIG. 2 *Taylor and Merchant's curves: Aggregate consolidation ratio U_{p+s} versus time t*

TABLE 2 TIME TO REACH U_{p+s} EQUAL TO 99.9% FROM TAYLOR AND MERCHANT'S THEORY IN A CLAY WITH, $\mu = 1.5 \times 10^{-7}\,s^{-1}$ AND R = 0.3

	Thickness (H) of double drained layer				
$c_v (m^2/s)$	2 cm	20 cm	2 m	10 m	20 m
4×10^{-7}	500 days	500 days	500 days	4,000 days	15,000 days
	1.37 year	1.37 year	1.37 year	11.0 years	41.1 years
10^{-8} (see	500 days	500 days	6000 days	150 000 days	600,000 days
Fig. 4)	1.37 year	1.37 year	16.4 years	411 years	1,644 years

Fig. 1). As a consequence, it might then be admitted that for all predictions based on Taylor and Merchant's theory with reliable soil parameters, with emphasis on a realistic value of c_v corresponding to the range of effective vertical stress (σ'_{vi} to σ'_{vf}) effectively applied, not a simple task when $\sigma'_{vi} < \sigma'_p < \sigma'_{vf}$, the real value of secondary compression will always be equal to or higher than the predicted one.

The authors built a field abacus for layer 6 using the parameters obtained from the oedometer tests in the virgin compression range but the field results plotted very close to the recompression line where the c_v value would be very different from the one used to build the abacus. Since, obviously, it would not be proper to use such an abacus in this situation, the authors were left with only one alternative to back-analyze the measured compression in layer 6, under area 3 and that was to try and adjust some Taylor and Merchant's curve to the measured curve determining the c_v value which would lead to a reasonable fit. As settlement occurs, the "pseudo final effective stress" which governs the consolidation process at any given time, that is "$\sigma_v - u_{hid}$" in layer 6 decreases with time due to embankment submersion. For their back-analysis, the authors took the decision to calculate the values of "$\sigma_v - u_{hid}$" acting in October 2009 in the middle of each of the 1 m thick sublayers considered for settlement calculations and the primary and secondary compressions which would occur if this "pseudo final effective stress" was maintained constant thereafter, that is the "pseudo total final compression" illustrated on Fig. 3, which, as shown, continuously decreases with time to finally reach the real final total compression, that is ΔH_t ($= \Delta H_p + \Delta H_s$) equal to 1.38 m. The calculated values in October 2009 are:

- pseudo final effective stress at mid-height of layer 6:
  $\sigma_v - u_{hid} = 179.2$ kPa,

- pseudo final primary compression (ΔH_p)': 20 cm,

- pseudo final secondary compression (ΔH_s)': 142 cm and

- pseudo final total compression (ΔH_t)': 162 cm.

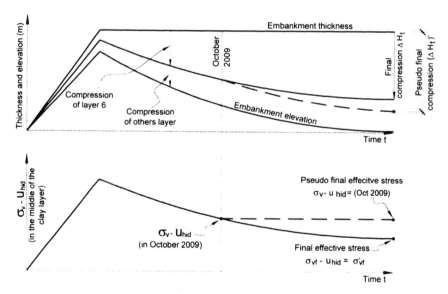

FIG. 3 *Evolution of the "pseudo final effective stress" in the middle of layer 6 and of the "pseudo final compression"*

It was immediately found out that no satisfying adjustment was to be obtained with only one constant value of c_v over the whole period. Fig. 4 shows a theoretical Taylor and Merchant's curve reasonably well fitted to the measured compression curve of layer 6 in the center of area 3.

This adjustment was achieved by dividing the total measurement period into three periods. As can be seen on Fig. 4, the values of c_v which led to the adjustment of the theoretical curve to the measured one are:

- for the first period: $c_v = 2.0 \times 10^{-7} \, m^2/s$,

- for the second period: $c_v = 5.0 \times 10^{-8} \, m^2/s$,

- for the third period: $c_v = 1.0 \times 10^{-8} \, m^2/s$

The ratio of the calculated primary compression in October 2009 to the "pseudo final primary compression" was found to be equal to 40%.

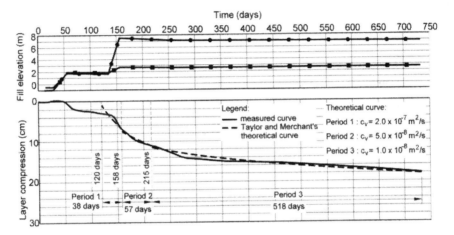

FIG. 4 *Theoretical Taylor and Merchant's curve reasonably well fitted to the measured compression curve of layer 6 under the center of area 3*

For the back-analysis of the embankment settlement data for layer 6 under the centers of areas 1 and 2 the same approach is not applicable due to the fact that the mathematical formulation presently available based on Taylor and Merchant's theory only encompasses the case of vertical drainage. In this respect, the authors are not aware of any published formulation of secondary consolidation associated with radial drainage.

It was then decided to compare the primary consolidation process of a layer 6 undergoing vertical compression through pure radial drainage with a mean radial drainage distance R equal to 0.68 m which corresponds to the drain spacing of 1.2 m of layer 6 under area 1, to the primary consolidation process through pure vertical drainage with a vertical drainage distance D to see if a D value could be found for which both processes would follow approximately the same course.

Since, for layer 6 the value of c_{h1} determined from CPT porepressure dissipation tests which corresponds to the recompression range according to Teh and Houlsby (1991) as shown by Rémy et al. (2010), is close to the oedometer test c_{v1} value, it has been assumed that the horizontal coefficient of consolidation c_h is equal to the vertical coefficient of consolidation c_v in both stress ranges.

For this comparison, the radial primary consolidation process was calculated using Barron (1948) formulas considering a horizontal/radial coefficient of consolidation $c_h = 1.5 \times 10^{-8}\,\text{m}^2/\text{s}$, an equivalent radius r_w of 3.25 cm for the 0.5 cm $\times$ 10 cm wick drain and a smear zone around each drain with a radius r_s equal to twice the drain equivalent radius, and a permeability coefficient equal to a fifth of the undisturbed clay permeability coefficient. The vertical primary consolidation process was calculated using Terzaghi and Frohlich (1936) formulation with a parametrical vertical drainage distance equal to D and $c_v = 1.5 \times 10^{-8}\,\text{m}^2/\text{s}$ (that is $c_v = c_h$). Fig. 5 shows the radial drainage curve for the above mentioned conditions and the vertical drainage curves for D values of 2.03 m and 1.81 m. As can be seen the vertical drainage curve for D = 2.03 m reaches the primary consolidation ratio of 50% at the same time as the radial drainage curve and the vertical drainage curve for D = 1.81 m reaches the primary consolidation ratio of 80% at the same time as the radial drainage curve. These D values, called the equivalent vertical drainage distances (Deq.), were also computed for the drain spacing of 2.4 m under area 2 and the results are summarized in Table 3.

TABLE 3 EQUIVALENT VERTICAL DRAINAGE DISTANCE DEQ.

Spacing of wick drains	Radial drainage distance	Deq. to attain U = 50% at the same time	Deq. to attain U = 80% at the same time
1.20 m[*]	R = 0.68 m	D eq. = 2.03 m	D eq. = 1.81 m
2.40 m[*]	R = 1.35 m	D eq. = 4.33 m	D eq. = 3.86 m

[*] considering in both cases the same smear zone

It is worth saying that these obtained D eq values are quite influenced by the radius and the coefficient of permeability of the smear zone.

It was then decided to use the following procedure to theoretically predict the course of compression of layer 6 under areas 1 and 2.

The course of the primary compression was calculated using Barron (1948) formulation considering the "pseudo final primary compression" $(\Delta H_p)'$, equal to 68 cm and 51 cm, respectively calculated as already mentioned, under the "pseudo final effective stresses", acting at mid height of layer 6 of 221.3 kPa and 221.9 kPa, respectively.

The course of the secondary compression was calculated using the formulas of Carvalho, isolating the secondary compression from primary compression, considering it to occur in a layer drained vertically with a drainage

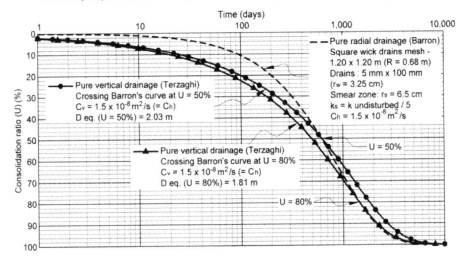

FIG. 5 *Comparison of primary consolidation ratio between a clay layer with vertical drainage distance D with no wick drain (pure vertical drainage) and a clay layer with a 1.2 m ×1.2 m mesh of wick drains (pure radial drainage) for $c_v = c_h$*

distance equal to 2.03 m and 4.33 m, respectively with the same "pseudo final primary compression" mentioned above and with the "pseudo final secondary compression" (ΔH_s)' calculated as already mentioned equal to 177 cm and 153 cm, respectively.

Finally, the primary and secondary compressions were added to get the course of the total compression.

It was immediately found out that no satisfying adjustment was to be obtained with only one constant value of c_v over the whole period.

Fig. 6 shows the theoretical curve obtained following the above described hybrid "primary Barron + pseudo secondary Taylor and Merchant procedure" reasonably well fitted to the measured compression curve of layer 6 under the center of area 1.

This adjustment was achieved by dividing the total measurement period into two periods. As can be seen on Fig. 6, the values of c_h (and c_v) which led to the adjustment of the theoretical curve to the measured one are:

· for the first period: $c_h = c_v = 5.0 \times 10^{-8} \, \text{m}^2/\text{s}$,

· for the second period: $c_h = c_v = 2.3 \times 10^{-8} \, \text{m}^2/\text{s}$,

The ratio of the calculated primary compression in October 2009 to the "pseudo final primary compression" was found to be equal to 72%.

It has to be pointed out that in the above analysis, the vertical drainage in layer 6 has not been taken into account for considering that since Deq equal to 2.03 m is only about one third of the vertical drainage distance equal to 5.5 m, it would not affect significantly the results of the back-analysis.

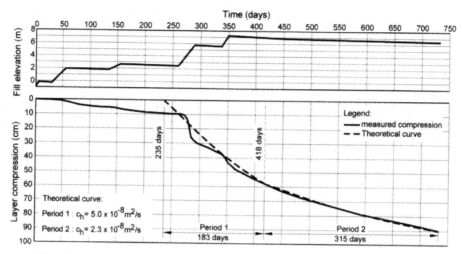

FIG. 6 *Theoretical curve reasonably well fitted to the measured compression curve of layer 6 under the center of area 1*

Fig. 7 shows a theoretical Taylor and Merchant's curve reasonably well fitted to the measured compression curve of layer 6 under area 2.

This adjustment was achieved by dividing the total measurement period into three periods. As can be seen on Fig. 7, the values of c_h (and c_v) which led to the adjustment of the theoretical curve to the measured one are: (in this case, due to the fact that differently from the two previous cases the whole first period ended before the final heightening of the embankment, the "final pseudo primary and secondary compressions" considered in the first period

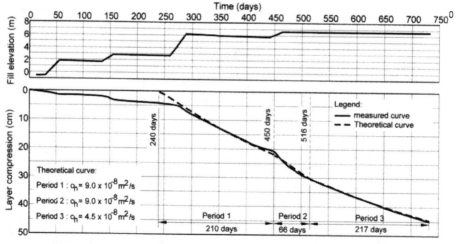

FIG. 7 *Theoretical curve reasonably well fitted to the measured compression curve of layer 6 under the center of area 2*

to establish the theoretical curve are the ones which are associated with the end of period 1, that is 34 cm and 152 cm, respectively, with a "pseudo final effective stress" of 205.9 kPa at mid height of layer 6):

- for the first period: $c_h = c_v = 9.0 \times 10^{-8}$ m^2/s,

- for the second period: $c_h = c_v = 9.0 \times 10^{-8}$ m^2/s,

- for the third period: $c_h = c_v = 4.5 \times 10^{-8}$ m^2/s

The ratio of the calculated primary compression in October 2009 to the "pseudo final primary compression" was found to be equal to 37%.

It has to be pointed out that in the above analysis, the vertical drainage in layer 6 has not been taken into account, although it would certainly have some influence on the consolidation process and on the obtained results, in this case leading to smaller values of $c_h = c_v$.

Table 4 summarizes the main results of the back-analysis of layer 6 compression, indicating for each area in the middle of the layer:

- the back-analysis method used as already mentioned,

- the initial vertical effective stress σ'_{vi},

- the "pseudo final vertical effective stress" "$\sigma_v - u_{hid}$" (Oct. 2009),

- the final vertical effective stress σ'_{vf},

- the preconsolidation stress σ'_p obtained from oedometer tests,

- the coefficient of consolidation for which the best fit was obtained between the theoretical curve and the measured compression curve, in each period: c_v (or c_h) during period 1, c_v (or c_h) during period 2, c_v (or c_h) during period 3,

- the calculated primary compression, ΔH_{p1}, and secondary compression, ΔH_{s1}, at the end of period 1,

- the calculated primary compression, ΔH_{p2}, and secondary compression, ΔH_{s2}, at the end of period 2,

- the calculated primary compression, ΔH_{p3}, and secondary compression, ΔH_{s3}, at the end of period 3,

- the calculated ratio of the primary compression calculated at the end of each period to the "pseudo final primary compression" $(\Delta H_p)'$. This ratio is an indication of the primary consolidation ratio reached at the end of each period.

TABLE 4 MAIN RESULTS OF BACK ANALYSIS OF LAYER 6 COMPRESSION

	Area 3	Area 1	Area 2
Drains	No	$1.2\,m \times 1.2m$	$2.4\,m \times 2.4\,m$
Method	Taylor and Merchant	"Hybrid primary Barron + secondary Taylor and Merchant"	
σ'_{vi}	81.4 kPa	81.2 kPa	86.6 kPa
σ_v - u_{hid} (in October 2009)	179.2 kPa	221.3 kPa	221.9 kPa
σ'_{vf}	160.6 kPa	201.6 kPa	198.3 kPa
σ'_p (lab)	187.3 kPa	187.1 kPa	192.5 kPa
Period 1			
c_v(or c_h)	$20 \times 10^{-8} m^2/s$	$5 \times 10^{-8} m^2/s$	$9 \times 10^{-8} m^2/s$*
ΔH_{p1}; ΔH_{s1}	4 cm; 3 cm	34 cm; 24 cm	9 cm; 13 cm
$\Delta H_{p1}/(\Delta H_p)'$	20%	50%	26%
Period 2			
c_v (or c_h)	$5 \times 10^{-8} m^2/s$	$2.3 \times 10^{-8} m^2/s$	$9 \times 10^{-8} m^2/s$*
ΔH_{p2}; ΔH_{s2}	6 cm; 5 cm	49 cm; 41 cm	13 cm; 17 cm
$\Delta H_{p2}/(\Delta H_p)'$	30%	72%	25%
Period 3			
c_v (or c_h)	$1 \times 10^{-8} m^2/s$		$4.5 \times 10^{-8} m^2/s$*
ΔH_{p3}; ΔH_{s3}	8 cm; 10 cm		19 cm; 26 cm
$\Delta H_{p3}/(\Delta H_p)'$	40%		37%

* keeping in mind that these c_v values would be lower if the influence of vertical drainage in layer 6 had been taken into account

Regarding the above back-analyzed c_h ($= c_v$) values, it has to be emphasized that when secondary compression and primary compression occur simultaneously, the volume of water expelled in this situation is larger than the volume of water which would be expelled if only primary consolidation took place. Since the initial pore pressure and gradient are not affected by the phenomenon of secondary consolidation, then the pore pressure dissipation has to be slower when secondary consolidation occurs simultaneously with primary consolidation than if it did not. This fact has been pointed out by Garlanger (1971) among others. Larsson et al (1997) can be quoted:

> At compression, the time dependence leads to a larger compression than that calculated from the compression moduli alone. A condition for this extra compression to occur is that the corresponding amount of water flows out of the soil. In turn, a condition for this to occur within the same period of time as the compression corresponding to the moduli alone is that a higher gradient exists in the pore water. The immediate effect of the creep tendency (or time effects) during a short time step is therefore an increase in pore pressure, whose size is determined by the pontential creep deformation and the compression modulus.

This is illustrated on the left side of Fig. 8. The right side of Fig. 8, shows the comparison of the measured settlement of an embankment soft clay with Larsson et al.'s prediction.

Taylor and Merchant's theory, although only stated in terms of void ratio, does not take into account the influence of the occurrence of secondary compression simultaneously with primary compression on the pore pressure, as clearly shown by the formulas of Carvalho, but this influence will necessarily lead to a field c_v (or c_h) value lower than the laboratory pure primary consolidation c_v (or c_h) value as clearly illustrated at the top of the left side of Fig. 8.

Figs. 9 and 10 show the back-analyzed values of c_v (= c_h) versus equivalent mean effective vertical stress σ'_v in the middle of layer 6 in areas 1, 2 and 3, respectively, with σ'_v calculated through the formula:

$$\sigma'_v = [(\sigma_v - u_{\text{hid}}) - \sigma'_{vi}] \times [\Delta H_{pi}/(\Delta H_p)']$$

It is well known that the yielding stress decreases when the strain rate decreases. It has, therefore, to be expected that the vertical effective stress at which the field c_v value passes from its higher value, in the recompression

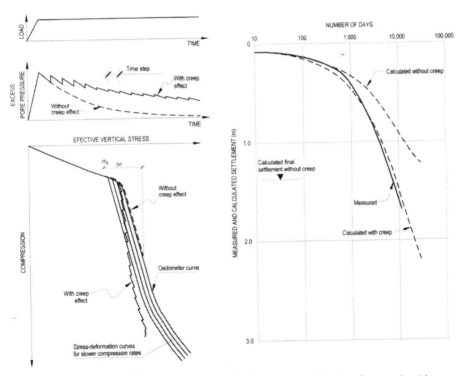

FIG. 8 *Influence of the occurrence of secondary compression simultaneously with primary compression apud Larsson et al. (1997)*

range, to its lower value, in the virgin compression range, is to be smaller than the σ'_p value determined in the oedometer tests.

It can be seen in Figs. 9 and 10 that the obtained results are in good agreement with the above observation and that the back-analyzed c_v ($= c_h$) field values are also in good agreement with the laboratory values in the transition range between the recompression range and the virgin compression range.

Close to the top and to the bottom of layer 6 under area 3, the local consolidation ratio is above 80% with σ'_v close to the "pseudo final effective stress $\sigma_v- u_{hid}$". At this stress level the c_v value is close to the small virgin compression range values. The very fast decrease of the c_v value under area 3 is, therefore, believed to be due to this "sealing" effect of the top and bottom sublayers.

It is also interesting to note that the c_v ($= c_h$) values under areas 2 and 3 also decrease towards the range of values of the virgin compression range. If, as advocated by various authors like Saye (2001) the driving of the wick drains had the effect of remoulding a large volume of soil around the drains up to the point of remoulding the whole soft clay volume in the case of closely spaced

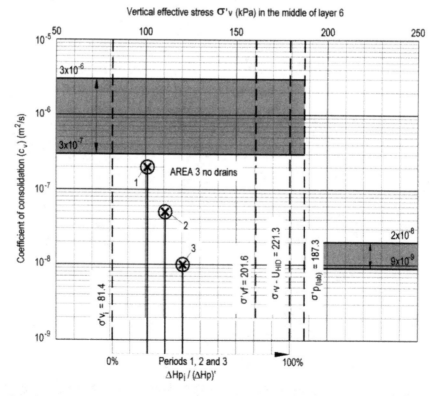

FIG. 9 *Back-analyzed values of $c_v(=c_h)$ versus equivalent mean effective vertical stress in the middle of layer 6 under area 3*

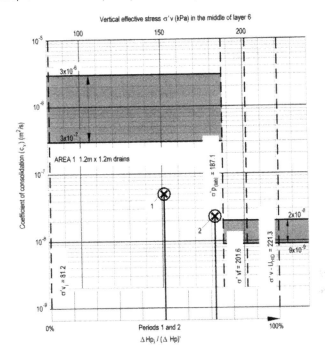

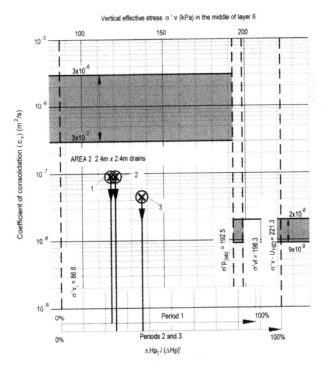

FIG. 10 *Back-analyzed values of $c_v(=c_h)$ versus equivalent mean effective vertical stress in the middle of layer 6 under areas 1 and 2*

drains as is the case under area 1, then the back-analyzed c_v (= c_h) values would not behave as shown in Fig. 9. These values would be much smaller even for lower effective stresses. It is also rather remarkable that the c_v (= c_h) values obtained from two totally distinct compression curve segments of layer 6, for period 1 and period 2 under area 2 for about the same value of σ'_v, are identical.

3   COMMENTS ON CURRENT METHOD USED TO ANALYZE EMBANKMENTS SETTLEMENTS

As mentioned in Rémy et al. (2010), the current method used to back-analyze embankments settlements is the Asaoka's (1978) method which considers that the secondary settlement only starts occurring after the end of the primary settlement. Table 5 shows the results of Asaoka's method applied during the period from 40 to 220 days using $\Delta t = 10$ days to a known layer compression (primary plus secondary) versus time curve of the Taylor and Merchant's type for $r = 0.5$, $c_v = 10^{-8}$ m^2/s and $\mu = 10^{-7}$ s^{-1}.

TABLE 5   RESULTS OF ASAOKA'S METHOD APPLIED TO A KNOWN LAYER COMPRESSION (PRIMARY + SEC-ONDARY) VERSUS TIME CURVE OF THE TAYLOR AND MERCHANT'S TYPE FOR R = 0.5, C$_v$ = 10^{-8} M^2/S AND μ = 10^{-7} S^{-1}

Layer thickness H (m)		2	8	20
	Predicted	0.28	0.29	0.44
Primary Compression ΔH_p (m)	Known	0.25	1.00	2.50
	Ratio	1.12	0.29	0.18
	Predicted	57%	55%	41%
Primary Consolidation Ratio U (%)	Known	49%	12%	5%
	Ratio	1.2	4.6	8.2
	Predicted	1.6	22	57
Consolidation coefficient c_v (10^{-8}m^2/s)	Known	1.0	1.0	1.0
	Ratio	1.6	22	57

As can be seen, the results obtained through Asaoka's method in this situation, a rather realistic one in the authors' eyes, are totally mistaken, leading to overestimate by a factor 22 the value of c_v in the case of an 8 m thick layer and by a factor of 57 for a 20 m thick layer. At the same time the final primary settlement is grossly underestimated, being only 29% and 18% of its known value for an 8 m thick layer and a 20 m thick layer respectively. The consolidation ratio values at the end of the reading period are grossly over-estimated. All these discrepancies are provoked by, and only by, the fact that the settlements occurring due to secondary compression are wrongly taken as part of the primary compression.

As already mentioned,

> when secondary compression and primary compression occur simultaneously, the volume of water expelled is larger than the volume of water which would be expelled if only primary consolidation took place. Since the initial pore pressure and gradient are not affected by the phenomenon of secondary consolidation, then the pore pressure dissipation has to be slower when secondary consolidation occurs simultaneously with primary consolidation than if it did not.

It means that in this case the course of pore pressure dissipation departs from Terzaghi's theory and that the Asaoka procedure applied to the pore pressure measurements as proposed by Orleach (1983) might also lead to erroneous results.

4 FINAL COMMENTS

Values of the compressibility parameters related to primary compression, that is: $C_c/(1+e_0)$, $C_r/(1+e_0)$ and the preconsolidation vertical stress, σ'_p, which are critically important for the design of embankments on soft clay foundations provided by the soils investigations firms are usually not trustworthy due to the poor quality of the"undisturbed" samples and of the laboratory tests. No compressibility parameter related to secondary compression is provided, neither it is called for by the designers who take for granted that secondary consolidation never finishes.

The values of the coefficient of primary consolidation, c_v, notably in the recompression range, yielded by oedometer tests usually also fail to be trustworthy for the same reasons. When reliable lab c_v values are available, much confusion is usually involved in the process of defining the design field c_v values from the lab values very often due to the widely spread creed that field values are in most cases up to 10 to 100 times higher than the lab values. Secondary consolidation is almost always considered to only start after primary consolidation comes to an end, to last forever and to evolve proportionally to the logarithm of time governed by the lab $C\alpha$ value independently of the thickness of the clay layer. Long term oedometer tests of the COPPE rheology group have shown that secondary compression is finite and ends after reaching a state equivalent to a constant OCR value for a given clay. These tests also showed that the oedometer $C\alpha$ value is not constant through the whole period of secondary consolidation.

Trial embankments back-analysis have repeatedly provided erroneous values of primary compressibility and consolidation parameters for being performed through Asaoka's method which does not take into account the simultaneous occurrence of primary and secondary consolidations.

High quality standard and special oedometer tests, such as the ones performed and presented by the authors, do achieve to provide trustworthy values of all primary and secondary compression and consolidation parameters of the tested soft clay layer. These tests results allow to build a reliable abacus which relates the strain and strain rate with the vertical effective stress applied to the laboratory specimen in the case of vertical compression with pure vertical drainage. However, contrary to the authors'expectation, it is not possible to directly use this abacus to back-analyze measured field settlements.

In the lack of better practical available methods, the Taylor and Merchant's formulation allows for reasonably reliable modeling of soft clay field behavior, taking into account the simultaneous occurrence of primary and secondary consolidations. The range of field values of the coefficient of consolidation obtained from the back-analysis of the settlements of the area with no drain and of the two areas with vertical drains of the Embraport pilot embankment, considering the final primary and secondary settlements calculated from the lab compressibility parameters, agrees remarkably well with the laboratory values. As expected, the transition of higher values of c_v (or c_h) in the recompression range to lower values of c_v (or c_h) in the virgin compression range occurs for values of the effective vertical stresses which are smaller in the field than in the laboratory.

The literature versing on the field behaviour of soft clay taking into account the undeniable simultaneous occurrence of primary and secondary compressions is very abundant. This is much more so than when Leroueil et al. (1985) wrote that "this abundant literature has modified neither the common practice based on the Terzaghi theory nor the way of thinking on clay behaviour". However, no practical tool has emerged from all this literature which can readily be used in common practice. This lack of tool has been, in the authors' understanding, the main hindrance to changes in common practice. This situation has much to do with the very little communication and lack of joint endeavour between practicing engineers and researchers.

ACKNOWLEDGEMENTS

The authors are grateful to EMBRAPORT in the person of Mr. Juvêncio Terra, for authorizing the use of the pilot embankment monitoring data to produce and publish this paper, and to Construtora Norberto Odebrecht in the person of Mr. Gilberto Gomes, who was always available to the authors to provide the instruments readings and all other relevant informations.

REFERENCES

Asaoka, A. (1978). *Observational procedure of settlement prediction*, Soils and foundations, Japanese Society of Soil Mechanics Foundation Engineering, v.18, n. 4, pp. 87-101.

Barron, R. A. (1948). Consolidation of fine-grained soils by drain wells, *Transactions*, ASCE, v. 113, p. 718-742.

Bjerrum, L. (1972). *Embankments on soft ground*, Performance of earth and earth supported structures, ASCE, pp. 1-54.

Carvalho, S. R. L. (1997). *Uma teoria de adensamento com compressão secundária*, PhD thesis, COPPE-UFRJ, Rio de Janeiro, RJ, Brazil (in Portuguese).

Christie, I. F. (1964). A re-appraisal of Merchant's contribution to the theory of consolidation, *Géotechnique*, vol. 14, n. 4, pp. 309-320.

Garlanger, J. E. (1971). *The consolidation of clays exhibiting creep under constant effective stress*, Norwegian Geotechnical Institute, internal report, n° 50302, Oslo.

Gibson, R. E. and Lo, K. Y. (1961). *A theory of consolidation for soils exhibiting secondary compression*, Publication n° 41, Norwegian Geotechnical Institute, pp. 1-16.

Larsson, R.; Bengtsson, P. E. and Eriksson, L. (1997). *Prediction of settlements of embankments on soft, fine-grained soils – Calculation of settlements and their course with time*, Swedish Geotechnical Institute, Information 13E, Linköping.

Leroueil, S., Kabbaj, M., Tavenas, F. and Bouchard, R. (1985). *Stress-strain rate relation for the compressibility of sensitive natural clays*, Geotechnique, vol. 35, n° 2, pp. 159-180.

Martins, I. S. M. (2005). *Algumas considerações sobre adensamento secundário*, Conference given at Clube de Engenharia, Rio de Janeiro, Brazil (Personal communication, in Portuguese).

Orleach, P. (1983). *Techniques to evaluate the field performance of vertical drains*. Master thesis, Massachusetts Institute of Technology.

Rémy J. P. P., Martins, I. S. M, Santa Maria, P. E. L., Aguiar, V. N. and Andrade, M. E. S. (2010), *The Embraport pilot embankment – primary and secondary consolidations of Santos soft clay with and without wick drains – Part 1*, In: Symposium on New Techniques for Design and Construction in Soft Clays, Guaruja, SP, Brazil.

Saye, S. R. (2001). *Assessment of soil disturbance by the installation of displacement sand drains and prefabricated vertical drains*, GSP n° 119, ASCE, EUA.

Taylor, D. W. and Merchant, W. (1940). A theory of clay consolidation accounting for secondary compression, *Journal of Mathematics and Physics*, vol.19, n.3, pp. 167-185.

Teh, C. I. and Houlsby, G. T. (1991). An analytical study of the cone penetration test in clay, *Géotechnique* 41, no. 1, pp. 17-34.

Terzaghi, K and Frohlich, O. K. (1936). *Theorie der Setzung von Tonschichten*, Franz Deuticke, Vienna (translated into French - Théorie des tassements des couches argileuses, Dunod – 1939, Paris)

Vieira, L. O. M. (1988) *Contribuição ao estudo de adensamento secundário.* M.Sc. thesis, COPPE/UFRJ, Rio de Janeiro, RJ, Brazil (in Portuguese).